Warbird Watcher's Guide to the Southern California Skies

Steve Smith

Edited by Bill Hample

SUNBELT PUBLICATIONS
San Diego, California

Warbird Watcher's Guide to the Southern California Skies

First edition 2002
Book design and composition by W. G. Hample & Associates
Edited by Bill Hample
Project management by Jennifer Redmond
Cover design by Hinge Branding and Design
Cover illustration by Russel Redmond
Printed in the United States of America

Sunbelt Publications, Inc.
P.O. Box 191126
San Diego, CA 92159-1126
(619) 258-4911 (619) 258-4916 fax
www.sunbeltpub.com

07 06 05 04 03 02 5 4 3 2 1

Library of Congress Cataloging-in-Publication Data

Smith, Steve D., 1961-
Warbird watcher's guide to the southern California skies / Steve D. Smith.-- 1st ed.
p. cm. -- (Sunbelt guidebooks and maps)
Includes index.
ISBN 0-932653-51-0
1. Airplanes, Military--United States--Handbooks, manuals, etc. 2. Airplanes--Recognition--California, Southern--Handbooks, manuals, etc. I. Title. II. Series.
UG1243 .S63 2002
623.7'46--dc21

20022005864

All photos by Steve Smith unless otherwise noted.

Dedication

To Deanne,

....for making me believe I could do all this!

And,

To Robin,

...who gave me all the expert advice on computers!

— Steve Smith

Contents

Foreword

Sunbelt publishes books about adventures in the natural history and cultural heritage of the Californias; we see *Warbird Watcher's Guide to the Southern California Skies* as an extension of this publishing mission. Styled like a field guide to bird identification (features in common include identification markings and physical specifications), this guidebook is a tool to understanding the vital role of military aviation in our region. Information on the history and capabilities of each aircraft evokes the adventure and thrill of hearing the growl of classic warbirds, the roar of modern jets, and the nightly news reports of squadrons that are fighting on distant frontiers of freedom.

As a Vietnam-era Naval aviator, I jumped at the opportunity to publish *Warbird Watcher's Guide*. Over the course of its production, the book was a catalyst to my recalling many grand memories of younger days: seeing waves of P-51 Mustangs returning from training flights; serving with high school buddies in the USAF Ground Observer Corps; air control duty aboard USS *John S. McCain* (DL-3); T-28 carrier qualification aboard the old USS *Lexington* (CVT-16); flying search and rescue missions from our home ship USS *Hornet* (CVS-12); serving with helo detachments aboard ships with stirring, historic names like *Kitty Hawk, Constellation, Oriskany, Forrestal*, and *Enterprise.*

The warbirds selected for this field guide are U.S. military aircraft which may currently be seen in the skies of southern California. As with their natural counterparts, some are solitary and rarely spotted, while others may fly in flocks and be commonly seen. Each has been or is a significant player in the stirring adventure story of the military forces who are dedicated to the defense of our land and way of life. We hope that *Warbird Watcher's Guide* will help you, the reader, to gain the knowledge that will enable participation in this adventure with more understanding and enjoyment.

Lowell Lindsay, Publisher

Preface

Today southern California abounds with military air facilities, air shows, aircraft assembly plants, aircraft support facilities, air museums, and a plethora of small, medium, and large airports. Thus, at almost any instant, military aircraft of the past or present (and in some cases, the future) are flying over some area in southern California.

These are the "warbirds," those marvelous flying machines which have been and are defending us. Identifying them can become both interesting and fascinating. It becomes interesting when you can identify these warbirds in the air. Seeing them later on the ground at a military open house, museum, or air show can become absolutely fascinating!

We have given only the most pertinent characteristics to aid in the recognition of each aircraft. We have also described only the basic or best-known models. If you wish to go further, there are many books and websites which will give you detailed histories, specifications, and other data on the many versions of each aircraft.

We have tried to provide accurate information. If you find we have overlooked important aircraft or aircraft features in our region, we welcome your comments via correspondence to the publisher.

Keep 'em flying!

Steve Smith, Author
Bill Hample, Editor
San Diego, California

Section 1
Fighters

Northrop F-5 E/F *Tiger II*

Service: U.S. Air Force; U.S. Marine Corps

Things to look for:

A. Nose comes to a very sharp point.
B. Single tail.
C. Two small engine intakes under cockpit.
D. Exhaust extends past tail.
E. Wings mounted low on fuselage.

Description:

The F-5 E/F *Tiger II* is a supersonic, single place, highly maneuverable air superiority fighter. It features low cost, ease of maintenance, and the ability to be used for various types of ground-support and aerial intercept missions.

The F-5 was developed from a 1954 U.S. Government request for a lightweight fighter. The F-5 first flew on 30 July 1959. The F-5 E/F *Tiger II* is an upgraded version of the basic F-5 *Freedom Fighter* and first flew in August 1972. The *Tiger II* has a larger wing span than the F-5 *Freedom Fighter*. More than 3000 F-5 fighters and T-38 *Talon* trainers have been produced for the United States and 25 foreign countries.

The F-5 is currently used as a dissimilar fighter trainer in this region. Only one Marine Squadron, VMFAT-401, stationed at MCAS Yuma, Arizona, flies the F-5 *Tiger II.*

Specifications:

Length	48 feet, 2.5 inches
Wing Span	26 feet, 8.5 inches
Height	13 feet, 2.5 inches
Crew	1
Maximum Speed	918 knots
Service Ceiling	51,800 feet
Empty Weight	9683 pounds
Maximum Take-off Weight	24,676 pounds

Armament: The *Tiger II* is currently unarmed as a trainer, however it can carry two *Sidewinder* air-to-air missiles at its wing tips and mixes of various other air-to-air and air-to-surface missiles, bombs, and unguided rockets on its external mounted pylons. It can also carry two 20 mm cannons.

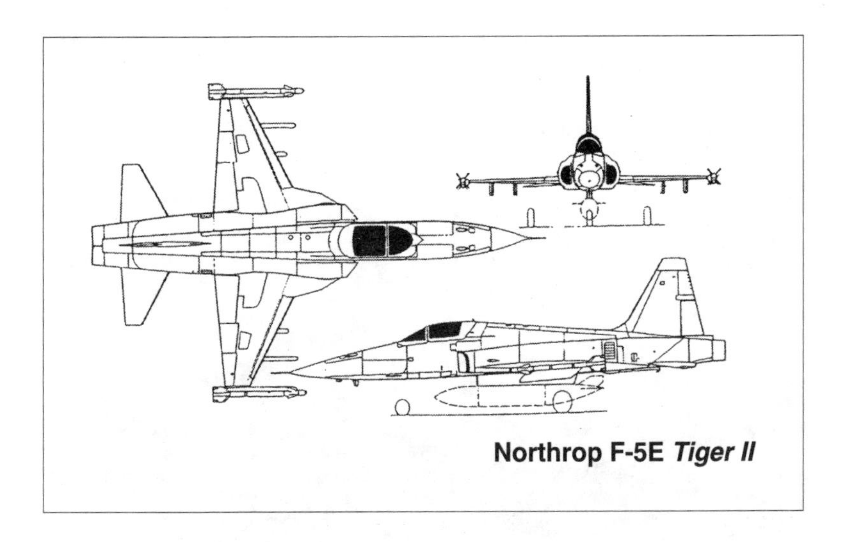

Northrop F-5E *Tiger II*

Grumman F-14 *Tomcat*

Service: U.S. Navy

Things to look for:

A. Twin vertical tails canted slightly outward.
B. Square jet intakes canted inwards at the top.
C. Jet engines are spread out.
D. Wings swing in and out.
E. Jet exhaust does not extend beyond vertical tail.

Description:

The Grumman F-14 *Tomcat* is a supersonic, twin-engine, variable sweep wing, two place fighter. The *Tomcat's* primary missions are air superiority, fleet air defense and precision strikes against ground targets.

The F-14 *Tomcat* was the latest in the historic Grumman "Cat" family. The *Tomcat* first flew in December 1970 and made its combat debut while deployed aboard the USS *Enterprise* (CVN 65) during operation "Eagle Pull" in September 1974. The F-14 provided Combat Air Patrol (CAP) missions, defending the fleet against manned bombers and other would be intruders. A total of 557 F-14As were built before production switched to the F-14A (Plus) redesignated F-14B a few years later. A total of 38 F-14Bs were built and 47 F-14As were brought up to F-14B standards.

Specifications:

Length	62 feet, 8 inches
Height	16 feet
Wing Span	64 feet, 1.5 inches @ 20-degree sweep (minimum); 38 feet 2.5 inches @ 58-degree sweep (maximum)
Crew (all models)	2
Maximum Speed	1078 knots
Service Ceiling	50,999-plus feet
Empty Weight	40,150 pounds
Maximum Take-off weight	74,150 pounds

Armament: The F-14 carries combinations of various air-to-air missiles including *Phoenix*, *Sparrow*, and *Sidewinder* on its external stations, and carries an internally mounted 20 mm Gatling cannon. Some aircraft are equipped with the LANTIRN targeting system that allows delivery of various laser-guided bombs for precision strikes in air-to-ground combat missions.

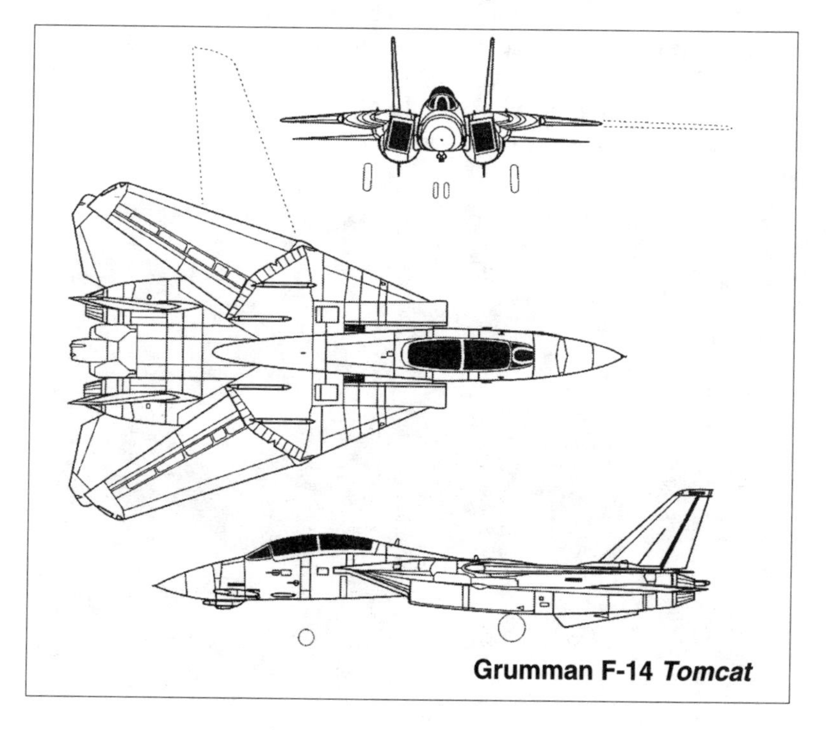

Grumman F-14 *Tomcat*

McDonnell Douglas (Boeing) F-15 *Eagle*

Service: U.S. Air Force

Things to look for:

A. Twin vertical tails.
B. Trailing edges of tails are straight.
C. Square engine intakes.
D. Jet exhausts are flush with the trailing edges of tails.
E. Two engines.
F. Engines are very close to each other.

Description:

This extremely maneuverable, all-weather tactical fighter is designed to permit the Air Force to gain and maintain air superiority in aerial combat. The F-15 has electronic systems and weaponry to detect, acquire, track and attack enemy aircraft while operating in friendly or enemy-controlled airspace.

The F-15 *Eagle* was announced as the winner of the F-X (Fighter-Experimental) competition for the USAF in 1969. The first F-15 made its maiden flight on 27 July 1972. A total of 905 F-15 A/B/C/D models were built. Originally built as an air-to-air fighter, the first F-15E *Strike Eagles* with air-to-air and air-to-ground capabilities were delivered to the Air Force during April 1988.

Specifications:

Length	63 feet, 9 inches
Wing Span	42 feet, 10 inches
Height	18 feet, 8 inches
Crew	1
Maximum Speed	Mach 2.5 at high altitude, 931 mph at low altitude
Service Ceiling	63,000 feet
Empty Weight	28,000 pounds
Maximum Weight	56,000 pounds

Armament: The *Eagle* can be armed with combinations of air-to-air weapons: *Sparrow* missiles or AMRAAM (advanced medium-range air-to-air missiles) on its lower fuselage corners, *Sidewinder* or AMRAAMs on two pylons under the wings, and an internal 20 mm Gatling cannon in the right wing root.

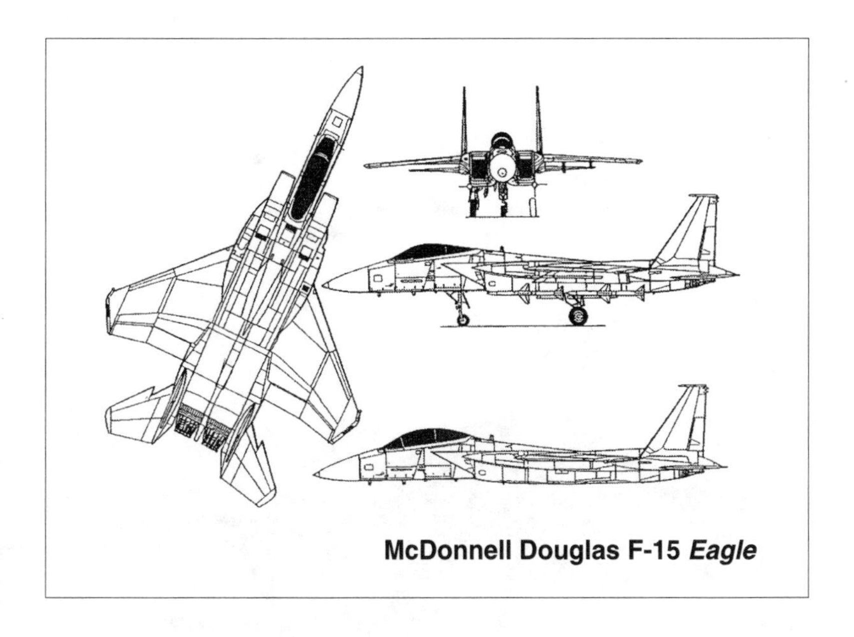

McDonnell Douglas F-15 *Eagle*

General Dynamics F-16 *Fighting Falcon*

Service: U.S. Air Force

Things to look for:

A. Engine intake located on bottom of aircraft.
B. Single tail.
C. Tail leans to the rear.
D. Horizontal tails' surface angle down.
E. Machine gun port under left side of cockpit.

Description:

A compact multi-role fighter, the F-16 *Fighting Falcon* is highly maneuverable and has proven itself both in air-to-air combat and air-to-surface attack. It provides a relatively low-cost, high-performance weapon system for the United States and allied nations.

The U.S. Air Force submitted a requirement for a light-weight fighter (LWF) during 1972. Several companies submitted designs. The General Dynamics F-16 and Northrop F-17 were selected for a fly-off. The F-16 first flew in January 1974. The fly-off between the F-16 and F-17 lasted until 1975 when the F-16 was selected. F-16s equip 25 U.S. Air National Guard units, four U.S. Air Force reserve units, 30 U.S. Air Force active duty units, and some 20 foreign air forces. More than 3900 F-16s have been built.

Specifications:

Length	49 feet, 4 inches
Wing Span	31 feet
Height	16 feet, 8.5 inches
Crew	1
Maximum Speed	1150 knots
Service Ceiling	60,000 feet, approximately
Empty Weight	18,600 pounds
Maximum Take-off Weight	- - -

Armament: The *Falcon* carries internally one 20 mm Gatling cannon. External stations can carry up to six air-to-air missiles, conventional air-to-air and air-to-surface munitions, and electronic countermeasure pods.

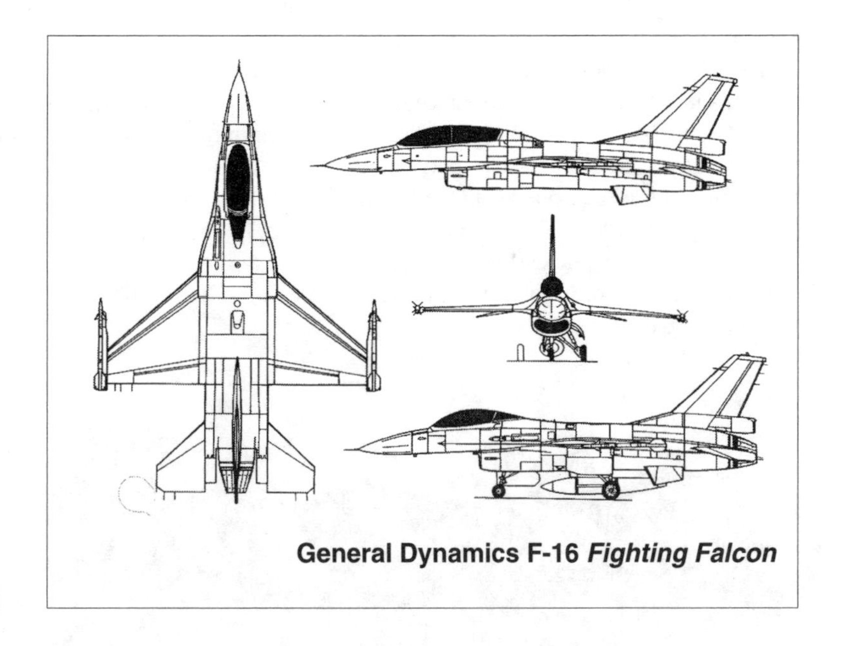

General Dynamics F-16 *Fighting Falcon*

McDonnell Douglas (Boeing) F/A-18 *Hornet*

Service: U.S. Navy; U.S. Marine Corps

Things to look for:

A. Twin vertical tails canted outward.
B. Jet exhaust extends well past the vertical tails.
C. Jet intakes are round on F/A-18A/B/C/D.
D. Jet intakes are square on the F/A-18E/F.
E. Engines are very close to each other.

Description:

The F/A-18 *Hornet*, an all-weather aircraft, is used as an attack aircraft as well as a fighter. In its fighter mode, the F/A-18 is used primarily as a fighter escort and for fleet air defense. In its attack mode, it is used for force projection, interdiction and close and deep air support.

The F/A-18 became operational with the U.S. Navy and the U.S. Marine Corps in 1983. The *Hornet* replaced the Navy's A-7 *Corsair II* and the Marine Corps F-4 *Phantom.* It assists the F-14 *Tomcat* in the fleet defense role. Currently a total of 41 active duty and eight reserve Navy and Marine Corps squadrons are equipped with the *Hornet*. During November 1995, the F/A-18E *Super Hornet* took to the air for the first time. The *Super Hornet* is 33 percent larger than the earlier model *Hornet* and serves only in the U.S. Navy. The Marine Corps has elected to wait for the Joint Strike Fighter to enter operational status.

Specifications:

Length	60 feet, 2.5 inches
Height	16 feet
Wing Span	44 feet, 10.75 inches; folded: 30 feet, 7.25 inches
Crew	1 (F/A-18E) 2 (F/A-18F)
Maximum Speed	1034 knots
Service Ceiling	50,000-plus feet
Empty Weight	30,500 pounds
Maximum Weight	66,000 pounds

Armament: The *Hornet* carries one 20 mm Gatling cannon. External payload can include combinations of various air-to-air missiles including *Phoenix*, *Sparrow*, *Sidewinder*, and AMRAAM; anti-ship missiles including *Harpoon*; the new Joint Stand-Off Weapon (JSOW); the new Joint Direct Attack Munition (JDAM); and various general purpose bombs, mines, and rockets.

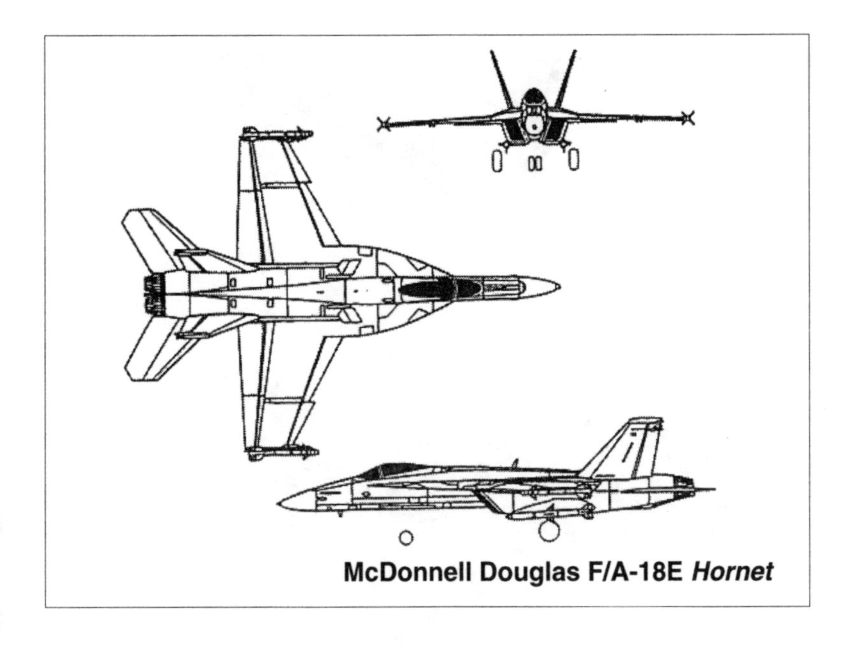

McDonnell Douglas F/A-18E *Hornet*

Section 2
Attack

Fairchild A-10 *Thunderbolt II*, aka *Warthog*

Service: U.S. Air Force

Things to look for:

A. Twin tail.
B. Two barrel-shaped engines on top of rear fuselage.
C. Barrel of Gatling gun extends under nose.
D. Cockpit is well forward of wings.
E. Landing gear housing located midway on wing.

Description:

This is the first Air Force aircraft specially designed for close air support of ground forces. It is a simple, effective, and survivable twin-engine jet aircraft that can be used against all ground targets, including tanks and other armored vehicles. Its wide combat radius and short takeoff and landing capability permit operations in and out of locations near front lines.

The A-10 *Thunderbolt II* beat out the Northrop A-9A (one is on display at March AFB) in a 1973 fly-off between the two aircraft. Production started in 1975 and the first squadrons were formed in 1976. The A-10 *Thunderbolt II* was built around the General Electric 30 mm Gatling gun and was designed to survive the battlefield environment. The engines are separated and are located high on the fuselage. The wings and tail provide extra protection for the engines. The main landing gear does not fully retract into the wheel wells. This provides protection in case of a belly landing. The pilot is protected by a titanium "bathtub." The A-10 *Thunderbolt II* flew more than 8500 combat missions during "Desert Storm."

Specifications:

Length	53 feet, 4 inches
Height	14 feet, 8 inches
Wing Span	57 feet, 6 inches
Crew	1
Maximum Speed	355 knots
Service Ceiling	Sea level, ground missions
Empty Weight	21,451 pounds
Maximum Weight	50,000 pounds

Armament: It carries one 30 mm Gatling cannon. Exernal payload includes combinations of mixed ordnance on eight under-wing and three under-fuselage pylon stations. Such weapons include 500-pound and 2000-pound bombs, incendiary cluster bombs, combined effects munitions, mine dispensing munitions, *Maverick* air-to-surface missiles, *Sidewinder* air-to-air missiles, 2.75-inch rockets, and laser-guided/electro-optically guided bombs.

Fairchild A-10 *Thunderbolt II*, aka *Warthog*

McDonnell Douglas AV-8B *Harrier II*

Service: U.S. Marine Corps

Things to look for:

A. Triangular tail.
B. Two outrigger wheels located in middle of wing.
C. Main landing gear in-line (bicycle) on fuselage.
D. Large half-circle intakes under cockpit on each side of fuselage.
E. Two exhaust nozzles located on each side of aircraft.

Description:

The AV-8B *Harrier II* is an upgraded model of the AV-8A *Harrier*. The *Harrier* is the first V/STOL (Vertical/Short Take-Off and Landing) aircraft to be operated by the United States. The mission of the *Harrier II* is to attack and destroy surface and air targets, to escort helicopters, and to conduct similar air operations. Combining tactical mobility, responsiveness, reduced operating cost and basing flexibility, both afloat and ashore, V/STOL aircraft are particularly well suited to the special combat and expeditionary requirements of the Marine Corps.

The *Harrier* was developed from the British-designed P1127 *Kestrel*. The *Kestrel* first flew on 19 December 1960. The Marine Corps received their first AV-8A *Harrier* in 1971, and retired the last "A" model during 1987. The newer AV-8B *Harrier II* entered service with the Marine Corps during 1985.

Specifications:

Length	47 feet, 9 inches
Height	11 feet, 8 inches
Wing Span	30 feet, 4 inches
Crew	1 (TAV-8B: 2)
Maximum Speed	574 knots
Service Ceiling	50,000 feet
Empty Weight	14,867 pounds
Maximum Take-off Weight	32,000 pounds standard

Armament: External payload can include combinations of mixed ordnance on seven external store stations, comprising six wing stations for *Sidewinder* air-to-air missiles and an assortment of air-to-surface weapons, including *Maverick* missiles. One centerline station can carry a pod of air-to-surface ordnance. A 25 mm Gatling cannon pod can also be mounted on the centerline.

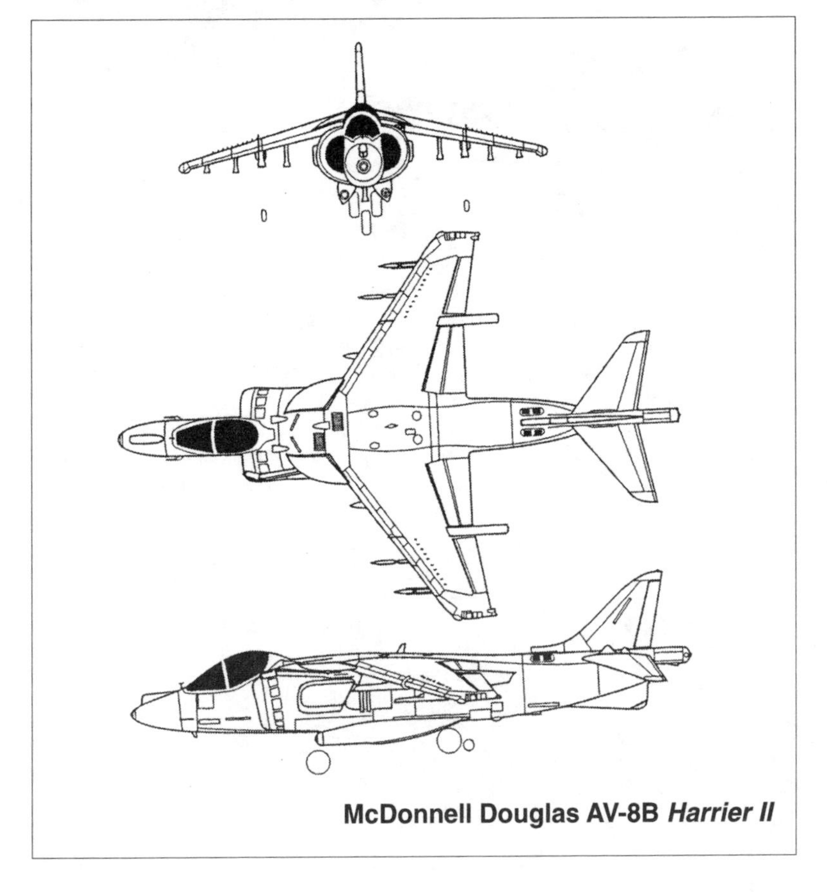

McDonnell Douglas AV-8B *Harrier II*

Section 3
Bombers

Rockwell B-1B *Lancer*

Service U.S. Air Force

Things to look for:

A. Four engines located under fuselage in pairs.
B. Canards under cockpit.
C. Horizontal tail surface located in middle of vertical tail.

Description:

The B-1 is the backbone of America's long-range bomber force providing massive and rapid delivery of precision and non-precision weapons against any potential adversary anywhere around the globe on short notice.

The B-1B's blended wing/body configuration, along with variable-geometry design and turbofan engines, combine to provide greater range and high speed with enhanced survivability.

The B-1A first flew on 23 December 1974. The B-1 program was canceled in 1977. The program was revived in October 1981 and featured an updated and larger aircraft, the B-1B. The B-1B's first flight was on 18 October 1984, with the first B-1B delivered to Dyess AFB in Texas in June 1985. The combat debut of the B-1B occurred on 17 December 1998 when two B-1Bs attacked the Al Kut barracks in Iraq. Currently 93 B-1B *Lancers* are in service with 80 *Lancers* available at anytime. Originally designed as a low-level penetration nuclear bomber, the B-1B's nuclear role is currently being reduced and modified to deliver conventional bombs instead of nuclear weapons.

Specifications:

Length	147 feet
Height	34 feet, 10 inches
Wing Span @ 15 sweep (minimum)	136 feet, 8.5 inches
Wing Span @ 67 sweep (maximum)	78 feet, 2.5 inches
Crew	4
Maximum Speed	Mach 1.25 (823 mph)
Empty Weight	192,000 pounds
Maximum Weight	477,000 pounds

Armament: Three internal weapons' bays can accommodate up to 84 Mk-82 general purpose bombs or Mk-62 naval mines, 30 CBU-87/89 cluster munitions or CBU-97 Sensor Fused Weapons, and up to 24 GBU-31 JDAM GPS guided bombs or Mk-84 general purpose bombs.

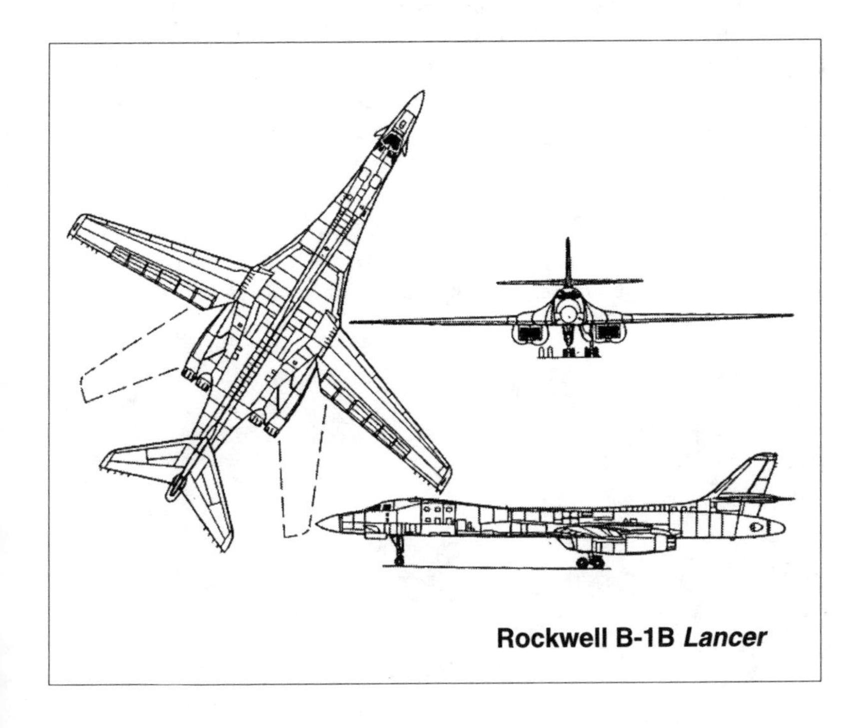

Rockwell B-1B *Lancer*

Northrop Grumman B-2 *Spirit*

Service: U.S. Air Force

Things to look for:

A. No tail.
B. Engine intakes on top of wing.
C. Trailing edge of wing is jagged.
D. No vertical surfaces.

Description:

This multi-role bomber is capable of delivering both conventional and nuclear munitions. Its low-observable, or "stealth," characteristics give it the unique ability to penetrate an enemy's most sophisticated defenses. These characteristics, its large payload, and extreme range (6000 nautical miles without refueling) give it important advantages over existing bombers.

The roots of the B-2 can be traced back to Jack Northrop's B-35 and B-49 flying wing bombers of the 1940s and 1950s. The B-2 *Spirit* has the same wing span as the B-35 and B-49, 172 feet. The B-2's first flight was on 17 July 1989 and it first entered service with the U.S. Air Force in December 1993. Two squadrons will be equipped with B-2s for a total of 21 aircraft. The combat debut of the B-2 occurred during Operation Allied Force in the Balkans in 1999, flying 35-hour non-stop missions from Whiteman AFB in Missouri to the Balkans and back. The 509th Squadron, the same unit that dropped the atomic bomb on Japan in World War II, currently is one of the squadrons flying the B-2.

Specifications:

Length	69 feet
Height	17 feet
Wing Span	172 feet
Crew	2
Maximum Speed	416 knots
Service Ceiling	50,000 feet
Empty Weight	125,000 pounds
Maximum Weight	371,300 pounds

Armament: The B-2 is intended to deliver combinations of gravity nuclear and conventional weapons, including precision-guided standoff weapons.

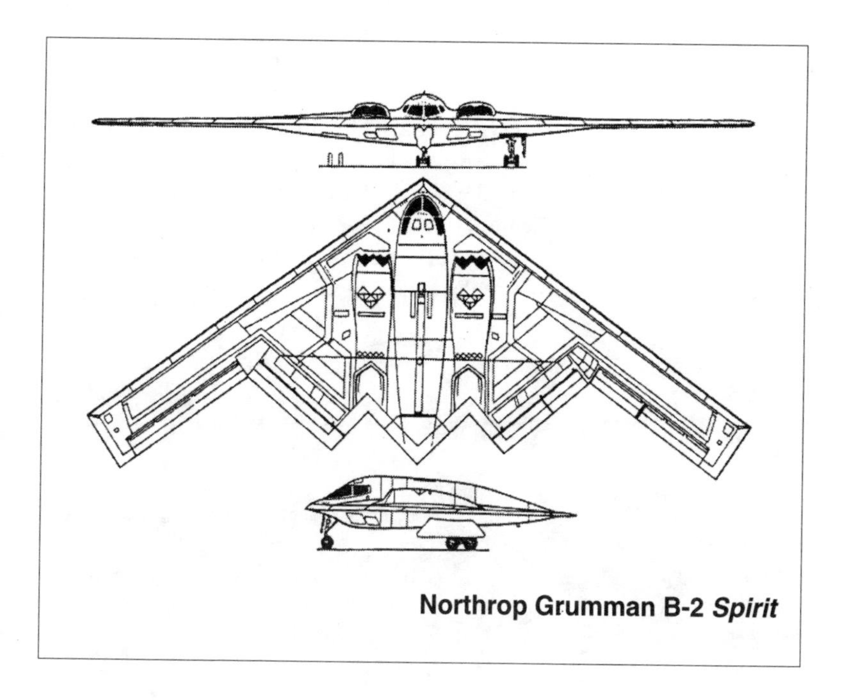

Northrop Grumman B-2 *Spirit*

Boeing B-52 *Stratofortress*

Service: U.S. Air Force

Things to look for:

A. Long slender fuselage.
B. Eight engines hung under the wings in pairs.
C. Outrigger wheels located on the tip of each wing.
D. Shoulder-mounted wings.

Description:

The B-52 is a long-range heavy bomber that can perform a variety of missions. It is capable of flying at high subsonic speeds at altitudes up to 50,000 feet. It can carry nuclear or precision guided conventional ordnance with worldwide precision navigation capability.

In a conventional conflict, the B-52 can perform strategic attack, air interdiction, and offensive counter-air and maritime operations. The use of aerial refueling gives the B-52 a range limited only by crew endurance. It has an unrefueled combat range in excess of 8800 miles.

The B-52 is the longest serving combat aircraft in history and is scheduled to remain in service for at least 20 more years. The B-52 first flew on 15 April 1952 and deliveries to the USAF started during 1955. The last of the B-52s rolled off the assembly line in 1962. The "H" model B-52 still serves today. Approximately 80 B-52Hs are available at any one time. The B-52 could last longer than its replacement, the B-1B *Lancer*.

Specifications:

Length	160 feet, 11 inches
Height	40 feet, 8 inches
Wing Span	185 feet
Crew	5
Maximum Speed	450 knots
Service Ceiling	55,000 feet
Empty Weight	170,000 pounds, approximately
Maximum Weight	505,000 pounds

Armament: The B-52 can carry approximately 70,000 pounds of mixed ordnance—bombs, mines, and missiles. Some models have been modified to carry air-launched cruise missiles, *Harpoon* anti-ship missiles, and other air-to-surface missiles.

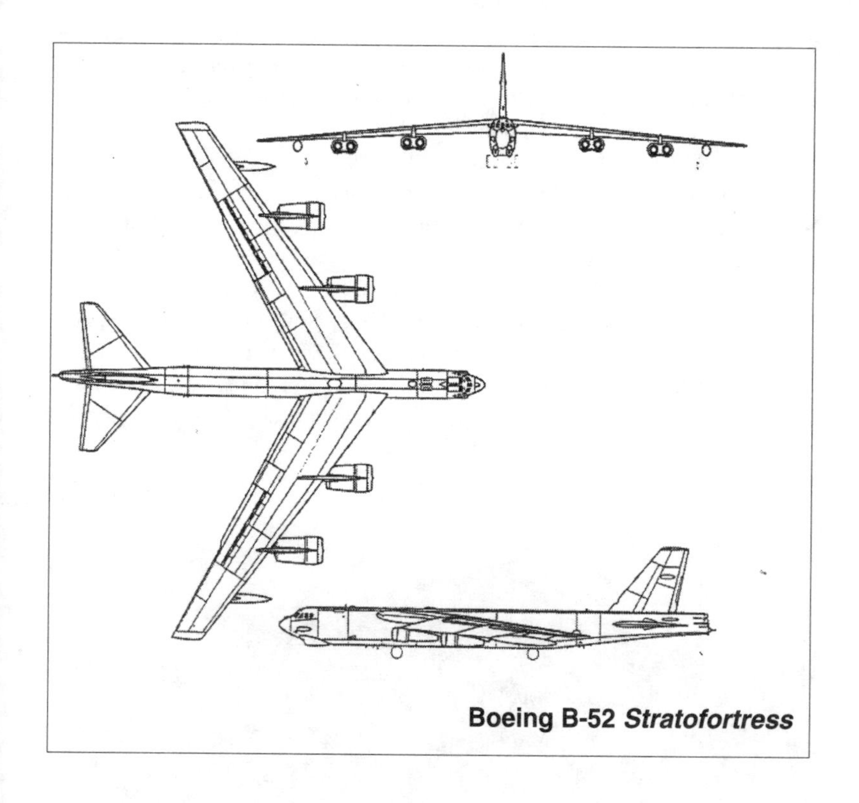

Boeing B-52 *Stratofortress*

Lockheed F-117 *Nighthawk*

Service: U.S. Air Force

Things to look for:

A. V-shaped tail.
B. Tail sweeps back.
C. Fuselage is extremely angular.
D. Aircraft is shaped like a triangle when viewed from underside.

Description:

The F-117A *Nighthawk*, or *Stealth Fighter*, is the world's first operational aircraft designed to exploit low-observable stealth technology. This precision-strike aircraft penetrates high-threat airspace and uses laser-guided weapons against critical targets. Air refuelable, it supports worldwide commitments and adds to the deterrent strength of U.S. military forces. The *Nighthawk* carries an "F" designator due to treaty agreements which limit the numbers of U.S. and Russian bombers.

The F-117 first flew on 18 June 1982. Not armed with guns, the F-117 carries only bombs, thus it qualifies more as an attack aircraft than a fighter aircraft. The first F-117 *Nighthawks* entered service during August 1982. These units were able to operate for almost 10 years in total secrecy. A total of 59 F-117s were delivered to the Air force. The last aircraft was delivered in 1990. The 8th and 9th Fighter squadrons operate the *Nighthawk* from Holloman AFB in New Mexico. Both units deployed to the Middle East for Desert Storm without a combat loss. During Operation Allied Force in Serbia, an F-117 *Nighthawk* was lost to a surface-to-air missile. The pilot ejected safely.

Specifications:

Length	65 feet, 11 inches
Height	12 feet, 5 inches
Wing Span	43 feet, 4 inches
Crew	1
Maximum Speed	Not available
Service Ceiling	Not available
Empty Weight	29,500 pounds
Maximum Weight	52,500 pounds

Armament: The aircraft can carry a combination of weapons in its internal weapons bay, including laser guided bombs, *Maverick* and HARM air-to-surface missiles, and other precision-guided weapons.

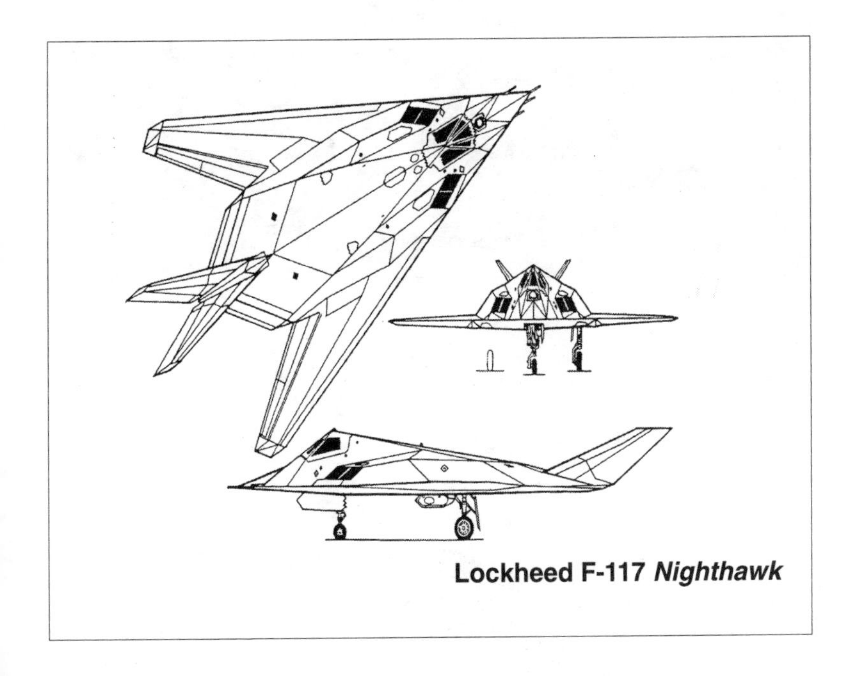

Lockheed F-117 *Nighthawk*

600
NAVY
NH

Section 4
ASW/Electronic Reconnaissance

Grumman E-2C *Hawkeye*

Service: U.S. Navy

Things to look for:

A. "Rotodome" radar housing over fuselage.
B. Large airscoop behind cockpit.
C. Four-bladed propeller.
D. Tail has four fins and twin rudders.
E. Round intakes located under propellers.
F. Twin turboprop engines.

Description:

The Grumman E-2C *Hawkeye* is the Navy's all-weather, carrier-based tactical warning and control system aircraft. It provides airborne early warning and command and control functions for the carrier battle group. Additional missions include surface surveillance coordination, strike and interceptor control, search and rescue guidance, and communications relay.

The E-2 *Hawkeye* series aircraft first entered service with the U.S. Navy in the 1964 and was initially designated W2F *Hawkeye*. The upgraded "C" model first flew 20 January 1971 and has been upgraded throughout the years. The E-2C *Hawkeye* can detect and track 2000 targets and direct 20 intercepts at the same time. The *Hawkeye* usually flies at 30,000 feet and sweeps an envelope of airspace of three million cubic square miles. The strange tail was designed to provide enough control and still fit on an aircraft carrier. The Rotodome on top of the aircraft is lowered when stored on board ship and raised when in flight.

Specifications:

Length	57 feet, 7 inches
Height	18 feet, 3.75 inches (rotodome in raised position); 16 feet, 5.5 inches (rotodome in lowered position)
Wing Span	Extended: 80 feet, 7 inches; folded: 29 feet, 4 inches
Crew	5
Maximum Speed	374 mph
Service Ceiling	31,000 feet
Empty Weight	38,200 pounds
Maximum Weight	51,933 pounds
Armament	None

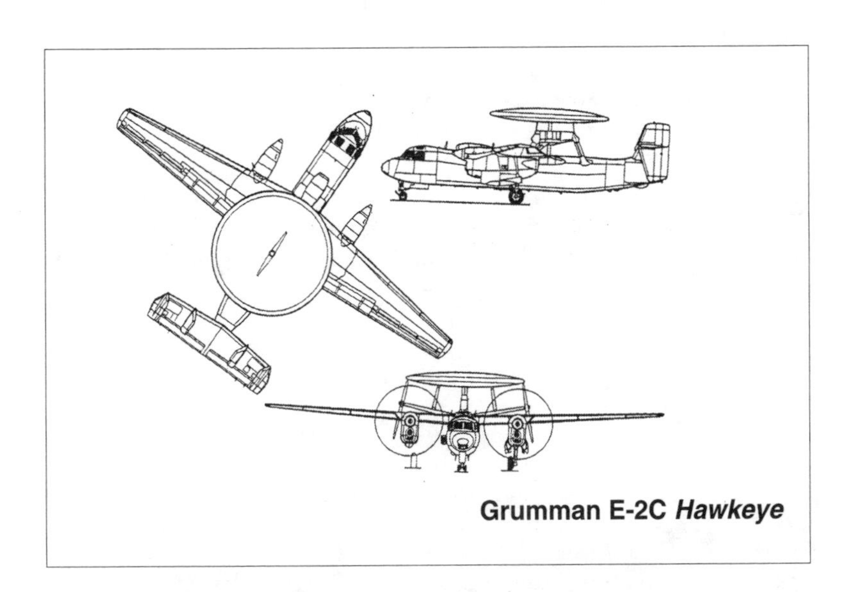

Grumman E-2C *Hawkeye*

Boeing E-3 *Sentry*

Service U.S. Air Force

Things to look for:

A. Low-mounted wing.
B. Engines located under wings on pylons.
C. Large disc radome located over fuselage.
D. Long cigar-shaped fuselage.
E. Vertical stabilizer is swept with curved fairing.

Description:

The E-3 *Sentry* is a flying headquarters providing Command, Control, and Communication (C3) needed by commanders of U.S., NATO, and other allied air defense forces. An all-weather surveillance aircraft, it carries Airborne Warning And Control Systems (AWACS) and is deployed near a combat zone to monitor aircraft and missiles and to direct friendly aircraft. As proven in Operation Allied Force, it is the premier air battle command and control aircraft in the world today.

The E-3's airframe is based on that of the Boeing 707, the same airframe used for the Boeing KC-135. The E-3 *Sentry* (EC-137D) first flew in February 1972. The first E-3 was delivered to the U.S. Air Force in March 1977.

The Rotodome, located over the fuselage, rotates six times per minute in operation, one-quarter turn per minute when not in operation, and weighs 3395 pounds. Thirty feet in diameter, it has a maximum depth of 6 feet and is capable of tracking 600 aircraft at once.

Specifications:

Length	152 feet, 11 inches
Wing Span	145 feet, 9 inches
Height	41 feet, 9 inches
Crew	20
Maximum Speed	530 mph
Empty Weight	171,950 pounds
Maximum Weight	325,000 pounds
Armament	None

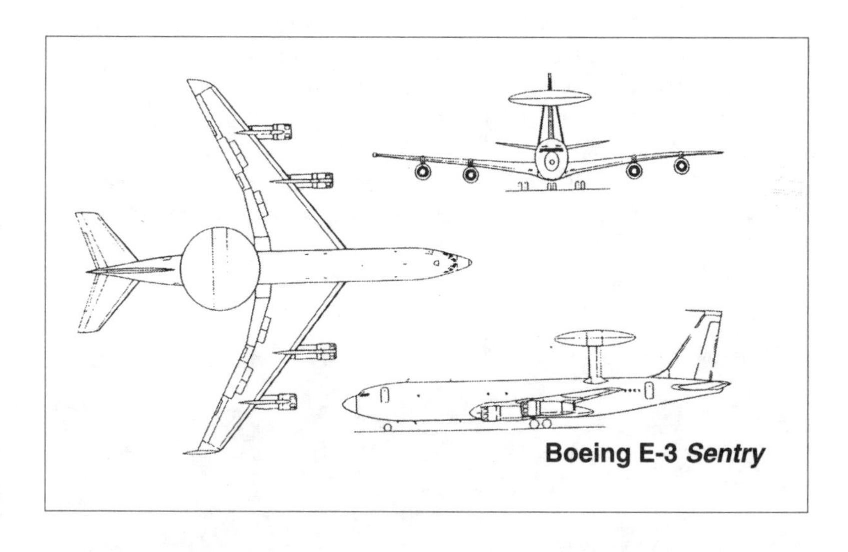

Boeing E-3 *Sentry*

Lockeed P-3 *Orion*

Service U.S. Navy

Things to look for:

A. Low-mounted wing.
B. Four turboprop engines mounted atop wings.
C. Boom mounted in tail.
D. Rounded vertical tail.

Description:

Lockheed's P-3 *Orion* is a long-range, four-engine, turboprop, anti-submarine patrol and maritime surveillance aircraft. Land based, the P-3C can carry a mixed payload of weapons internally and on wing pylons.

The prototype P-3 *Orion* first flew 25 November 1959. The P-3 *Orion* is based on the Lockheed L-188 *Electra* airliner. The first seven P-3A *Orions* entered U.S. Navy service during July 1962 as P3V-1 models. A total of 157 P-3A models were built. The P-3B model went into production during 1964. The P-3B had more powerful engines and updated sensors. The last P-3B was built in 1969. The P-3B was replaced by the P-3C version which entered service with VP-30 in 1969. The last P-3C *Orion* was delivered on 17 April 1990. The P-3C has been upgraded several times to increase its overall capability. The P-3C flew 369 combat missions for a total of 3787 hours during Desert Shield/Desert Storm. The P-3C *Orion* is also built by Kawasaki for the Japanese Maritime Self Defense Force.

Specifications:

Length	116 feet, 10 inches
Height	33 feet, 8.5 inches
Wing Span	99 feet, 8 inches
Crew	11
Maximum Speed	473 mph
Service Ceiling	28,300 feet
Empty Weight	61,490 pounds
Maximum Weight	142,000 pounds

Armament: Combinations of mixed ordnance include *Harpoon* and SLAM cruise missiles, *Maverick* air-to-surface missiles, torpedoes, rockets, mines, depth bombs, and various special weapons.

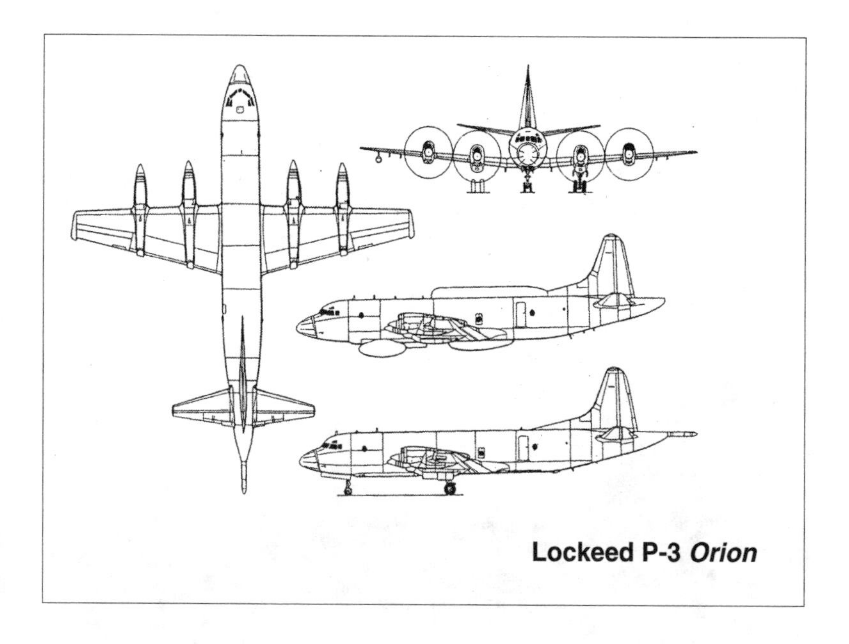

Lockeed P-3 *Orion*

Lockheed S-3B *Viking*

Service U.S. Navy

Things to look for:

A. Engines located in pods under wings.
B. Tall tail.
C. Wings mounted high on fuselage.
D. Tail begins very close to trailing edge of wings.
E. Flanked seat cockpit forward of wing.

Description:

This jet aircraft is used to hunt and destroy enemy submarines and provide surveillance of surface shipping. A carrier-based, subsonic, all-weather, long-range, multi-mission aircraft, it operates primarily with carrier battle groups in anti-submarine warfare zones. It carries automated weapon systems and is capable of extended missions with in-flight refueling. The ES-3 version is fitted for electronic warfare and reconnaissance.

The S-3 *Viking* first flew on 21 January 1972. The S-3 replaced the S-2F *Tracker* as the Navy's anti-submarine aircraft. The last S-3A *Viking* rolled off the assembly line in 1978. The first of 119 version S-3s, as modified to the S-3B configuration, flew during September 1984. The S-3 *Viking* will be continually upgraded until replaced by the planned Common Support Aircraft in 2015.

Specifications:

Length	49 feet, 5 inches
Height	22 feet, 9 inches; folded: 15 feet, 3 inches
Wing Span	Extended: 68 feet, 8 inches; folded: 29 feet, 6 inches
Crew	4
Maximum Speed	450 knots
Service Ceiling	40,000 feet
Empty Weight	26,650 pounds
Maximum Weight	52, 540 pounds

Armament: Combinations of mixed ordnance include *Harpoon* cruise missiles, *Maverick* and SLAM air-to-surface missiles, torpedoes, rockets, and bombs.

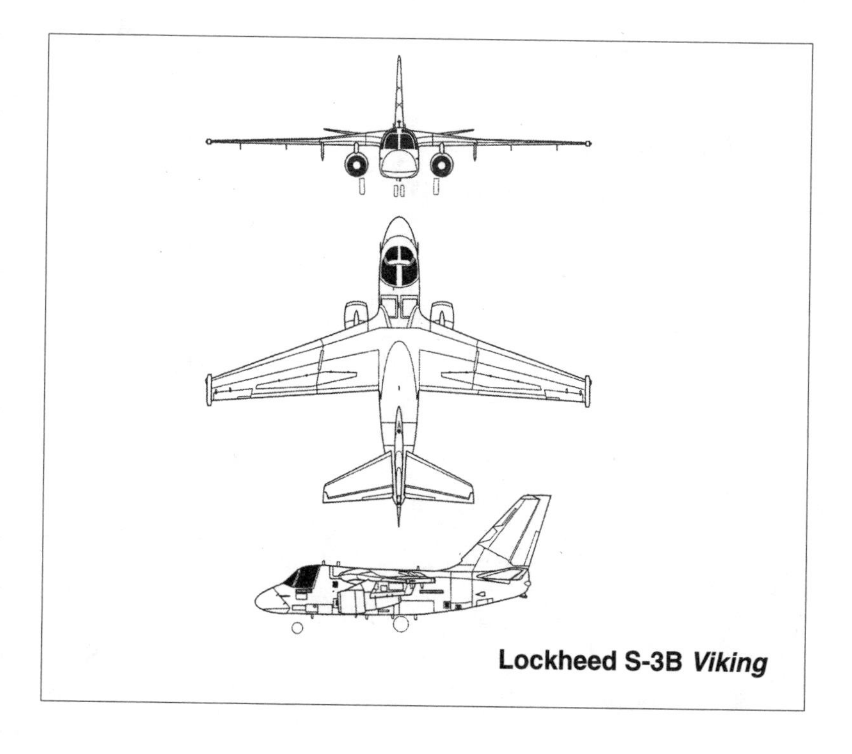

Lockheed S-3B *Viking*

Section 5
Cargo Aircraft

Grumman C-2 *Greyhound*

Service: U.S. Navy

Things to look for:

A. Tail has four fins and twin rudders.
B. One turboprop on each wing.
C. Wing mounted high on fuselage.
D. Very short nose.
E. Four-bladed propeller.

Description:

Powered by two T-56 turboprop engines, this cargo aircraft is designed to land on and provide critical logistics support to aircraft carriers. Its primary mission is Carrier On-Board Delivery (COD). Priority cargo such as jet engines can be transported from shore to ship in a matter of hours. The aircraft's folding wings, as well as an on-board auxiliary power unit for engine starting and ground power self-sufficiency in remote areas, provide an operational versatility found in no other cargo aircraft.

The C-2 *Greyhound* first flew on 18 November 1964. Nineteen aircraft were built between 1965 and 1968. Thirty- nine additional aircraft were built between 1982 and 1985. The C-2 *Greyhound* can trace its roots to the Grumman E-2C *Hawkeye*. The fuselage of the C-2 is much wider than the E-2C's and the landing gear has been beefed up for heavier landing loads due to its COD role. The C-2 *Greyhound* is capable of carrying 39 passengers, or 20 litter patients with four attendants, or 18,000 pounds of cargo.

Specifications:

Length	57 feet, 10 inches
Wing Span	Extended: 80 feet, 7 inches; folded: 29 feet, 4 inches
Height	16 feet, 10.5 inches
Crew	3
Maximum Speed	357 mph
Service Ceiling	33,500 feet
Empty Weight	36,346 pounds
Maximum Weight	57,000 pounds

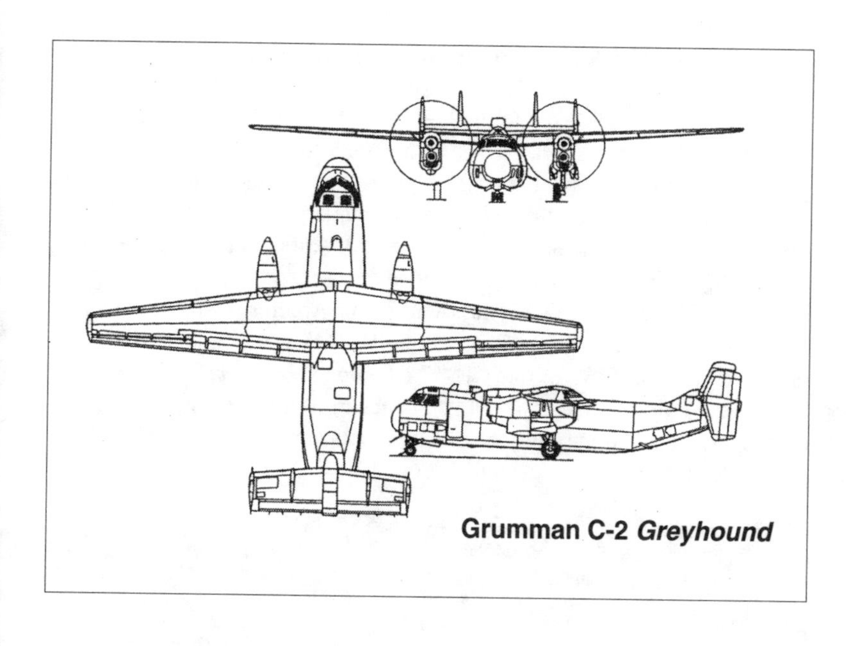

Grumman C-2 *Greyhound*

Lockheed C-5A/B *Galaxy*

Service: U.S. Air Force

Things to look for:

A. Tee tail.
B. Two engines mounted on pylons under each wing.
C. Wings mounted high on fuselage.
D. Bulbous nose.
E. Lower rear surface of fuselage slants upward.

Description:

With its tremendous payload capability, the C-5 provides the Air Mobility Command intertheater airlift in support of United States national defense. The C-5, the C-17 *Globemaster III* and the C-141 *Starlifter* are partners of AMC's strategic airlift concept. The aircraft carry fully equipped combat-ready military units to any point in the world on short notice then provide field support required to help sustain the fighting force.

The C-5A *Galaxy* first flew on 30 June 1968. The first C-5A production model was delivered to the U.S. Air Force on 17 December 1969. The last was delivered in May 1973. A shortfall of heavy lift capabilities in the 1980s required a reopening of the C-5 production line. Fifty new C-5B *Galaxys* were built and delivered between January 1986 and April 1989. The C-5B is basically identical to the C-5A. Modifications to the landing gear and flight control systems were added.

The C-5 is capable of carrying two M-1A1 Abrams main battle tanks, four M-551 Sheridan light tanks plus one HMMVW (Hummer), 10 light assault vehicles (LAVS), or one CH-47 *Chinook* helicopter. During Desert Shield/Desert Storm the C-5 flew 42 percent cargo and 18.6 percent passenger missions. There is currently no planned replacement for the C-5 and it is projected to remain in service until 2010.

Specifications:

Length	247 feet, 10 inches
Wing Span	222 feet, 8.5 inches
Height	65 feet, 1.5 inches
Crew	6
Maximum Speed	463 mph
Service Ceiling	35,750 feet
Empty Weight	374,000 pounds
Maximum Weight	837,000 pounds

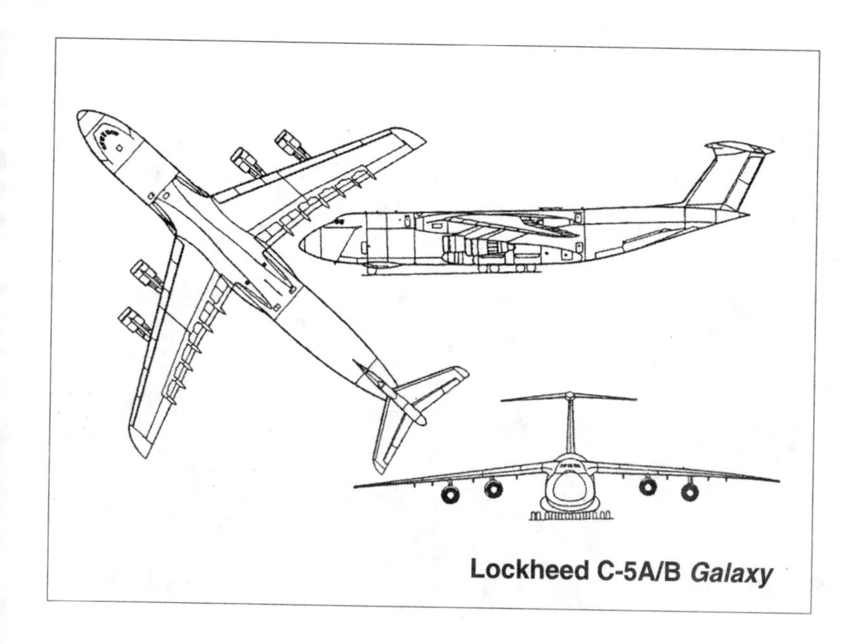

Lockheed C-5A/B *Galaxy*

Beech C-12 *Huron*

Service: All four U.S. military branches

Things to look for:

A. Tee tail.
B. Turboprop engine.
C. Three-blade propeller.
D. Round windows along fuselage.
E. Wings mounted in lower fuselage.
F. Wings pass through engine nacelles.

Description:

The C-12 *Huron* is a twin-engine logistics aircraft that carries passengers and cargo between military installations. Its cabin can readily accommodate cargo, passengers, or both. It is also equipped to accept litter patients in medical evacuation missions.

The C-12 *Huron* is a military version of the Beech *Super King Air*. The prototype *Super King Air* first flew on 27 October 1972. The C-12 *Huron* was procured as an-off-the shelf staff transport. All four branches of the U.S. military have acquired the C-12 *Huron*. The first C-12s were delivered to the U.S. Army in July 1975. Some models of the C-12 have been modified for electronic and intelligence gathering missions.

Specifications:

Length	43 feet, 9 inches
Wing Span	54 feet, 6 inches
Height	15 feet
Crew	2
Maximum Speed	299 mph
Service Ceiling	31,000 feet
Empty Weight	7315 pounds
Maximum Weight	12,500 pounds

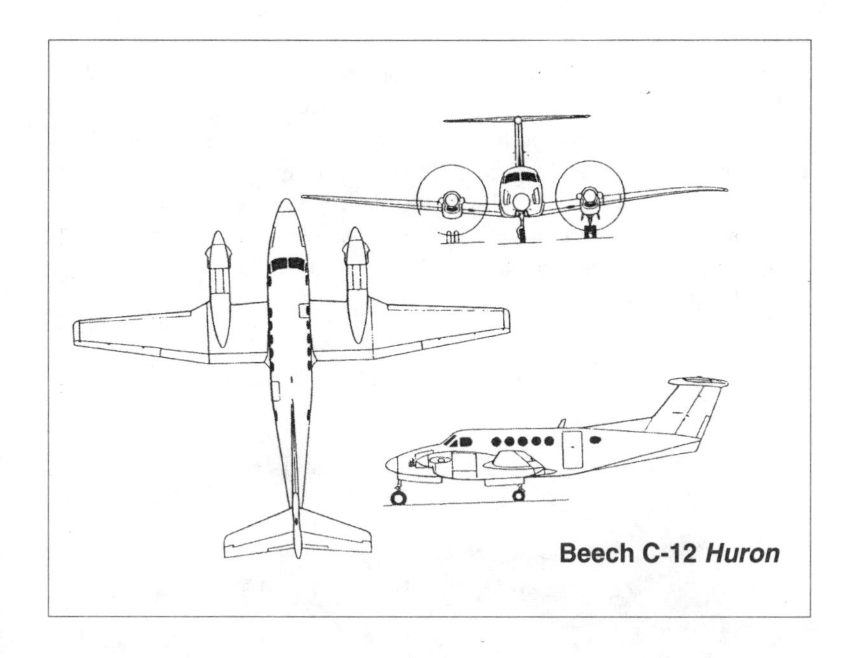

Beech C-12 *Huron*

Douglas C-17 *Globemaster III*

Service: **U.S. Air Force**

Things to look for:

A. Winglets on wing tips.
B. Tee tail.
C. Two engines on each wing.
D. Wings mounted high on fuselage.
E. Housing for landing gear located on side of fuselage.
F. Lower rear surface of fuselage slants upward.

Description:

This is the newest, most flexible cargo aircraft to enter the airlift force. It is capable of rapid strategic delivery of troops and all types of cargo to main operating bases or directly to forward bases in the deployment area.

The C-17 *Globemaster III* first flew on 15 September 1991. The C-17 *Globemaster III* is the first transport aircraft to have a HUD or Heads-Up-Display. The C-17 is also the first transport aircraft to have an ergonomically designed cockpit and a quadruple redundant fly-by-wire control system. The cargo area is equipped with a pallatized load/unload system that can be operated by one person. The C-17 *Globemaster III* can carry 102 troops, 48 litter patients, three AH-64 Apache attack helicopters, or air droppable platforms of up to 110,000 pounds.

Specifications:

Length	174 feet
Wing Span	171 feet, 3 inches
Height	55 feet, 1 inch
Crew	3
Maximum Speed	450 knots
Service Ceiling	45,000 feet
Empty Weight	269,000 pounds
Maximum Weight	580,000 pounds

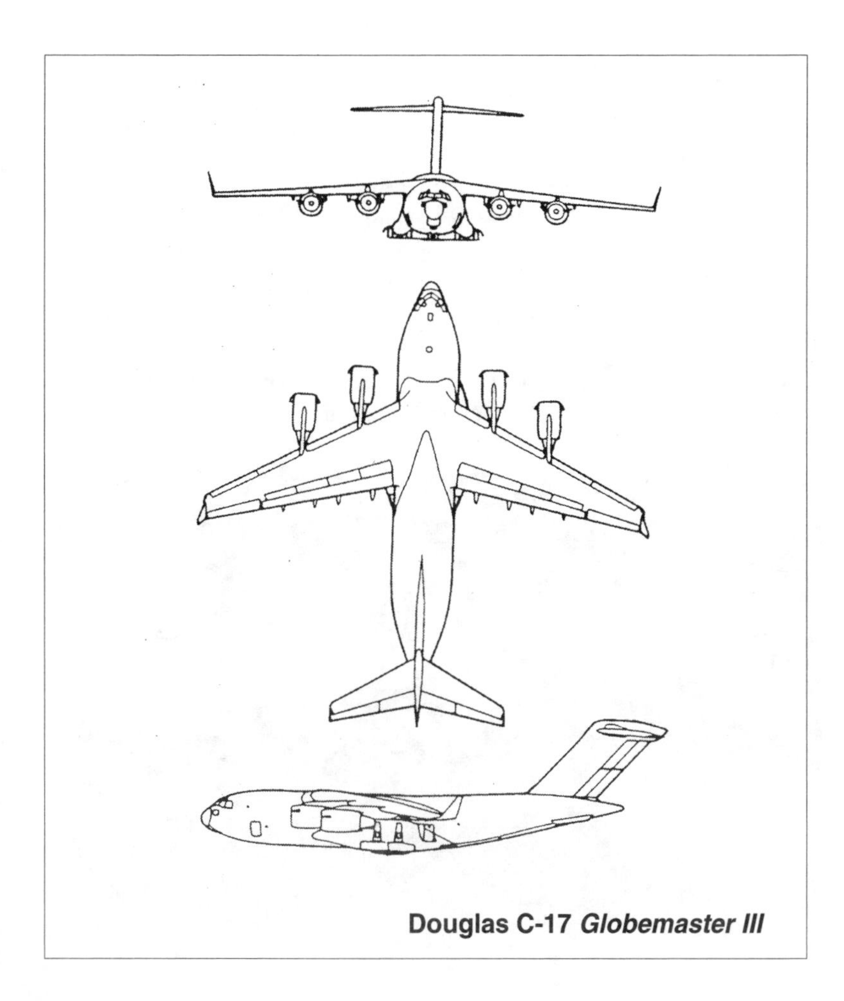

Douglas C-17 *Globemaster III*

Lockheed C-130 *Hercules*

Service: **U.S. Air Force; U.S. Coast Guard; U.S. Marine Corps; U.S. Navy**

Things to look for:

A. Four turboprop engines.
B. Triangular tail.
C. Wings mounted high on fuselage.
D. Landing gear located in pods on sides of fuselage.

Description:

The C-130 *Hercules* primarily performs the tactical portion of the airlift mission. The aircraft is capable of operating from rough dirt strips and is the prime transport for air dropping troops and equipment into hostile areas.

The prototype *Hercules* made its first flight 23 August 1954. Deliveries to U.S. Air Force units began during December 1956. The *Hercules* fulfills numerous roles, such as weather reconnaissance (WC-130), air-to-air refueling (KC-130), gunship (AC- 130), and drone director (DC-130). The LC-130 is ski-equipped for operations in the Antarctic. New models of the *Hercules* are continuing to appear, such as the C-130J with its more powerful engines and increased capabilities. The *Hercules* is probably one of the most successful aircraft of all time for its mission.

Specifications:

Length	97 feet, 9 inches
Wing Span	132 feet, 7 inches
Height	38 feet, 3 inches
Crew	5
Maximum Speed	330 knots (380 mph)
Service Ceiling	25,000 feet
Empty Weight	79,981 pounds
Maximum Weight	175,000 pounds
Armament	None (with exception of AC-130 gunship)

Lockheed C-130 *Hercules*

Lockheed C-141 *Starlifter*

Service: U.S. Air Force

Things to look for:

A. Tee tail.
B. Lower rear surface of fuselage slants upward.
C. Wings mounted high on fuselage.
D. Four engines mounted on pylons under wing.
E. Long slender fuselage.
F. A "hump" over the cockpit.

Description:

The C-141 *Starlifter* is the "workhorse" of the Air Mobility Command. It fulfills the vast spectrum of airlift requirements through its ability to airlift combat forces over long distances and deliver those forces and their equipment either by air, land, or airdrop. It can resupply forces and transport the sick and wounded from the hostile area to advanced medical facilities.

The C-141A first flew on 17 December 1963. It entered service with the U.S. Air Force in October 1964 and started operations in Vietnam in April 1965. The fuselage is equipped with paratroop doors and clamshell doors with a ramp to permit straight-in loading and off-loading. During the 1970s, C-141As were modified as C-141Bs. Major modifications consisted of plugs installed forward and aft of the wing, thus stretching the fuselage an additional 23 feet, 4 inches. Also added was an in-flight refueling receptacle located in a "hump" over the cockpit section. The C-141B first flew on 24 March 1977. The C-141B can carry 166 passengers in passenger seats, or 205 passengers in canvas seats, or 168 paratroopers. It can also carry 103 litter patients, five HMMWV (Hummers), one AH-1 Cobra, or one Sheridan tank.

Specifications:

Length	168 feet, 3 inches
Wing Span	159 feet, 11 inches
Height	39 feet, 3 inches
Crew	5
Maximum Speed	455 mph
Service Ceiling	41,500 feet
Empty Weight	144,000 pounds
Maximum Weight	343,000 pounds

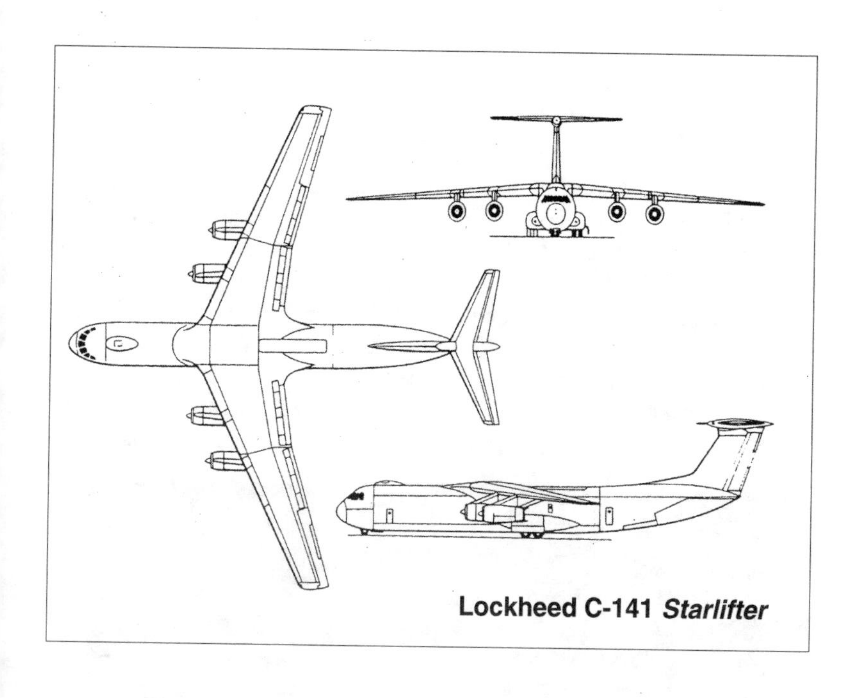

Lockheed C-141 *Starlifter*

McDonnell Douglas KC-10 *Extender*

Service: U.S. Air Force

Things to look for:

A. Two engines on pylons under each wing, third engine mounted in vertical tail.
B. Refueling boom located at rear of fuselage.
C. Large cargo door located between cockpit and leading edge of wing, left side of fuselage.
D. Extra set of wheels located in center of fuselage between main landing gear.
E. Large cargo door located behind cockpit and in front of wing.
F. Extra set of wheels located in center of fuselage between main landing gear.

Description:

The KC-10 is an Air Mobility Command advanced tanker and cargo aircraft designed to provide increased global mobility for U.S. armed forces. Although the KC-l0's primary mission is aerial refueling, it can combine the tasks of a tanker and cargo aircraft by refueling fighters and simultaneously carry the fighter support personnel and equipment on overseas deployments.

The KC-10 first flew on 12 July 1980. It is a version of an off-the-shelf commercial DC-10-30CF. The DC-10-30CF and KC-10 retain an 88 percent commonality in systems and parts.The KC-10 features military avionics, satellite communications, aerial refueling boom, hose and drogue system, and its own refueling receptacle. The aircraft entered service in March 1981; the last of 60 was delivered 29 November 1988. Delivery fuel totaling 18,125 gallons is carried in seven fuel bladders in the lower cargo bay.

Specifications:

Length	181 feet, 7 inches
Wing Span	155 feet, 4 inches
Height	58 feet, 1 inch
Maximum Speed	610 mph
Service Ceiling	33,400 feet
Empty Weight	244,630 pounds
Maximum Weight	590,000 pounds

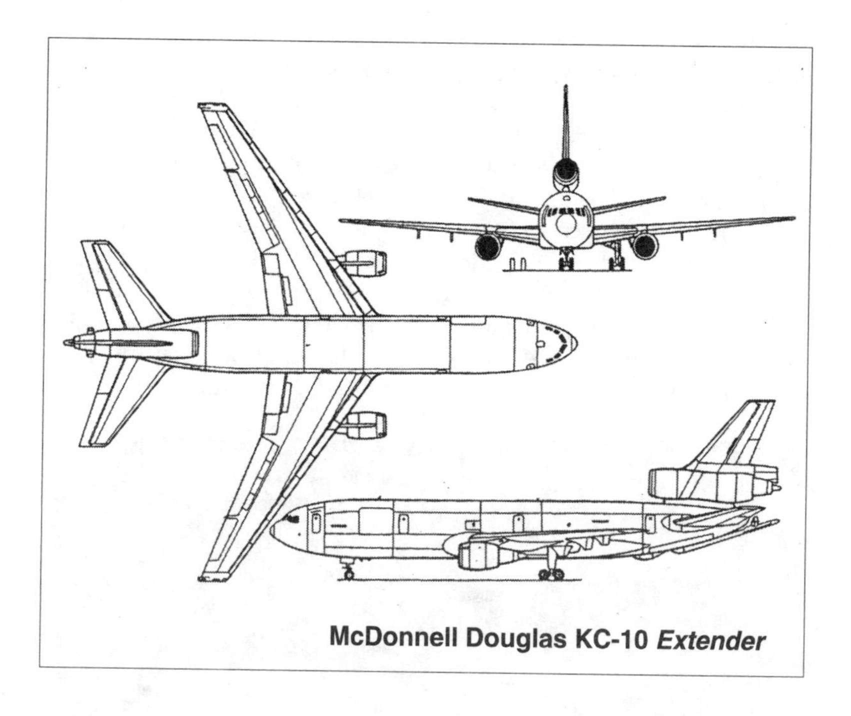

McDonnell Douglas KC-10 *Extender*

Boeing KC-135 *Stratotanker*

Service: U.S. Air Force

Things to look for:

A. Slender fuselage.
B. Very few windows on sides of fuselage.
C. Four jet engines located on pylons under wings.
D. Swept-back tail.
E. Refueling boom located under tail.

Description:

The KC-135 *Stratotanker's* principal mission is air refueling. This unique asset greatly enhances the USAF's capability to accomplish its primary missions of Global Reach and Global Power. It also provides aerial refueling support to Air Force, Navy and Marine Corps aircraft as well as aircraft of allied nations. Nearly all internal fuel can be pumped through the tanker's flying boom, the KC-135 's primary fuel transfer method. An operator stationed in the rear of the plane controls the boom. A cargo deck above the refueling system can hold a mixed load of passengers and cargo.

The KC-135 is based on Boeing's 707 airframe. The 707 first flew on 15 July 1954. The first KC-135 flew in August 1956. The first of the new KC-135s was delivered to the U.S. Air Force 28 June 1957. The KC-135 has gone through several engine upgrades over its service life. Fifty-six KC-135 were modified to the KC-135Q configuration to carry the special JP-7 fuel for the Mach 3 reconnaissance aircraft, the SR-71 *Blackbird.*

Specifications:

Length	136 feet, 3 inches
Wing Span	130 feet, 10 inches
Height	41 feet, 8 inches
Crew	3
Maximum Speed	600 mph
Service Ceiling	50,000 feet
Empty Weight	102,300 pounds
Maximum Weight	322,500 pounds

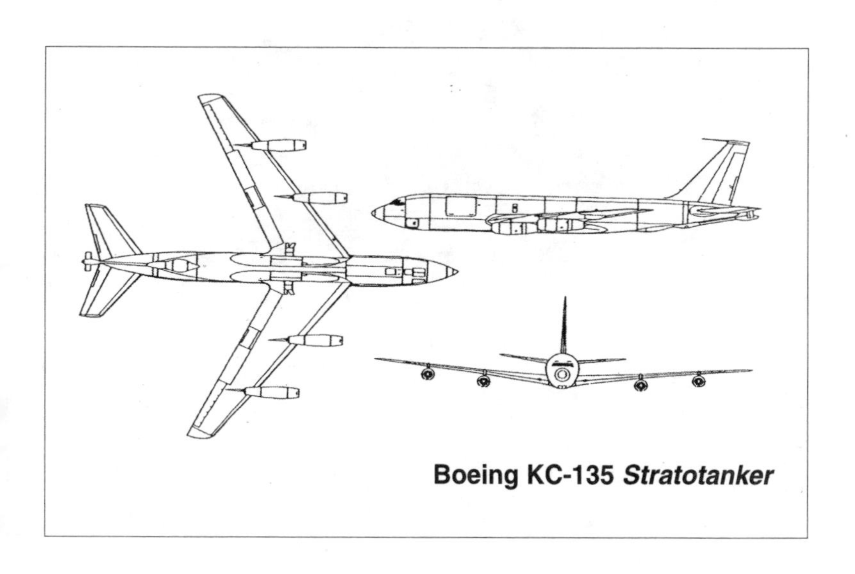

Boeing KC-135 *Stratotanker*

Section 6
Trainers

Beech T-34 *Mentor*

Service: U.S. Navy; U.S. Marine Corps

Things to look for:

A. Large glass canopy.
B. Large exhaust pipes on side of engine cowling.
C. Squared-off vertical stabilizer.

Description:

The tandem seat T-34 *Mentor* was used by both the U.S. Air Force and the U.S. Navy as their primary trainer. The USAF version was replaced in the late 1950s by the Cessna T-37 jet trainer. The Navy version, upgraded to a turboprop 400 hp engine, is still in use by the Navy as a basic trainer.

The T-34C *Mentor* first flew on 21 September 1973 and was developed from the Beech T-34 A/B *Mentor* which made its maiden flight in 1945. The *Mentor* is used in basic and intermediate pilot training, target spotting, and pilot proficiency checks. Currently more than 300 *Mentors* serve the Navy. Two serve in the Marine Corps with VMFAT-101 (Marine Fighter Attack Squadron, Training-101), the Fleet Replacement Squadron (FRS), at MCAS Miramar, San Diego, California.

Specifications:

Length	28 feet, 8.5 inches
Height	9 feet, 11 inches
Wing Span	33 feet, 5 inches
Crew	2
Maximum Speed	223 knots
Service Ceiling	30,000 feet
Empty Weight	2960 pounds
Maximum Weight	4300 pounds

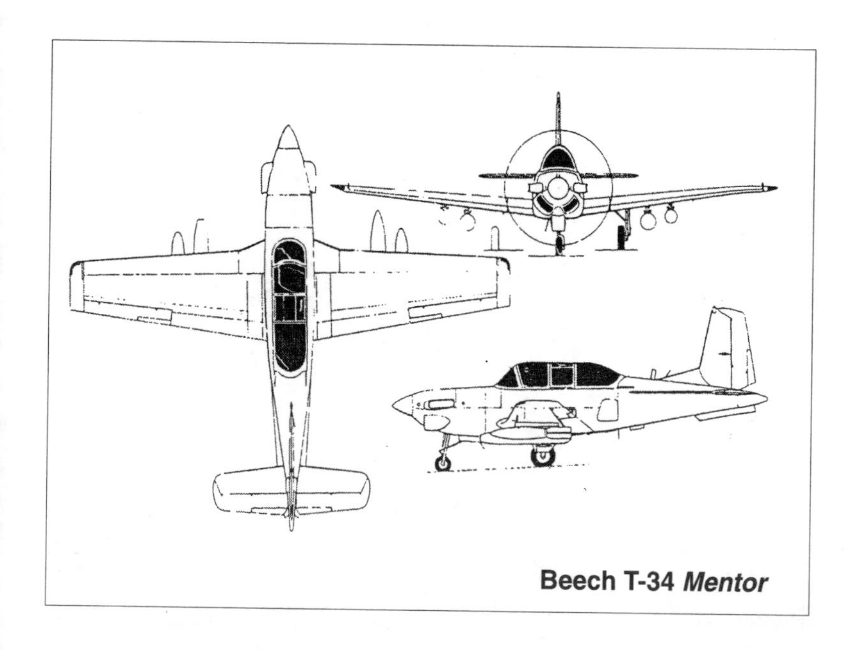

Beech T-34 *Mentor*

Northrop T-38 *Talon*

Service: U.S. Air Force

Things to look for:

A. Needle nose.
B. Tandem seat cockpit forward of the wing.
C. Long cockpit canopy.
D. Wing mounted low on fuselage.
E. Round engine intakes located at roots of wings.
F. Engines located next to each other, tail cones extend beyond tail.
G. Wing and tail are tapered, squared at the tips.

Description:

This twin-engine, high-altitude supersonic jet trainer is used in a variety of roles because of its design, economy of operations, ease of maintenance, high performance, and exceptional safety record. The instructor and student sit in tandem on rocket-powered ejection seats in a pressurized, air-conditioned cockpit.

The T-38 *Talon* first flew on 10 April 1959. The T-38 is a spin-off from the F-5 series fighter. Production of the T-38 started during 1961 and more than 1100 were completed. Production ended in 1972. The first of the T-38 *Talons* went to the 3510th Flying Training Wing at Randolph AFB, Texas. Nearly every Air Force pilot, from 1961 to 1993, has flown the T-38 *Talon*. The USAF Thunderbirds flew the T-38 from 1978 to 1984. The T-38, with upgrades and modifications, is projected to be in service until 2010.

Specifications:

Length	46 feet, 4.5 inches
Wing Span	25 feet, 3 inches
Height	12 feet, 10.5 inches
Crew	2
Maximum Speed	627 mph
Service Ceiling	53,600 feet
Empty Weight	7174 pounds
Maximum Weight	12,959 pounds

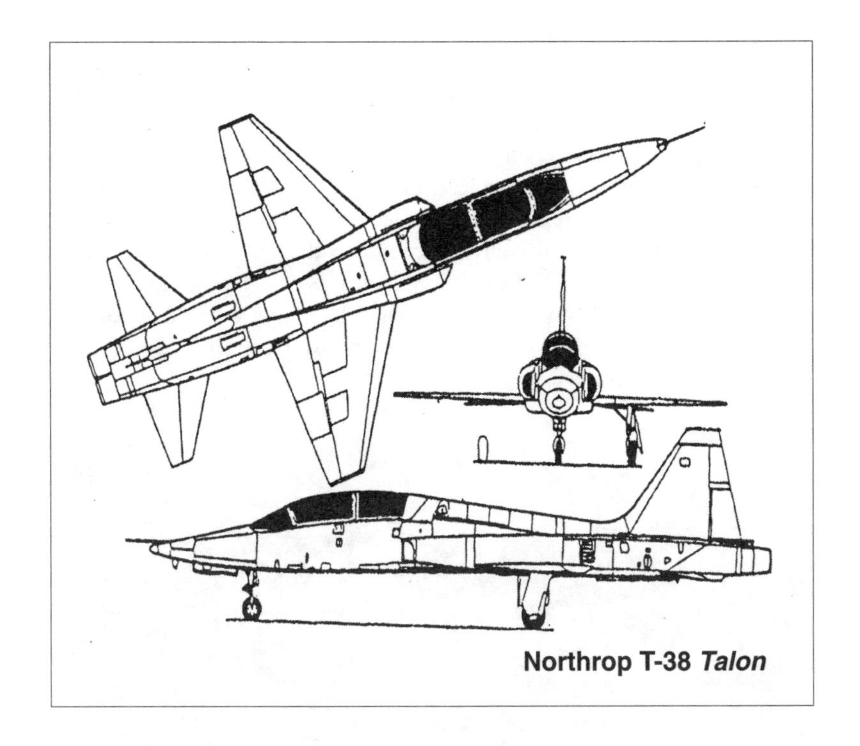

Northrop T-38 *Talon*

McDonnell Douglas T-45 *Goshawk*

Service: U.S. Navy

Things to look for:

A. Tandem seat cockpit.
B. Cockpit canopy has downward slope.
C. Wings mounted low on fuselage.
D. Horizontal tail surfaces slant downward.
E. Small jet intakes at wing roots and slightly forward of wing.

Description:

The T-45A *Goshawk* is a tandem-seat, carrier-capable jet trainer. It is used for intermediate and advanced portions of the Navy/Marine Corps pilot training program for jet carrier aviation and tactical strike missions.

Originally designed and built by British Aerospace for the RAF, the *Hawk* made its first flight on 21 August 1971. The U.S. Navy decided in 1981 to use a modified version of the British *Hawk* as its new trainer. A number of modifications were made to the original *Hawk* airframe. The British single air brake was replaced by two air brakes located on each side of the fuselage. The airframe was strengthened to withstand carrier operations. The single nose wheel of the British version was also replaced by a dual wheel nose gear. The Navy adopted the name *Goshawk* to avoid confusion with the U.S. Army's *Hawk* missile. The first McDonnell Douglas T-45 *Goshawk* flew on 16 December 1991.

Specifications:

Length	39 feet, 4 inches
Wing Span	30 feet, 10 inches
Height	13 feet, 6 inches
Crew	2
Maximum Speed	560 knots
Service Ceiling	18,800 feet
Empty Weight	9394 pounds
Maximum Weight	13,500 pounds

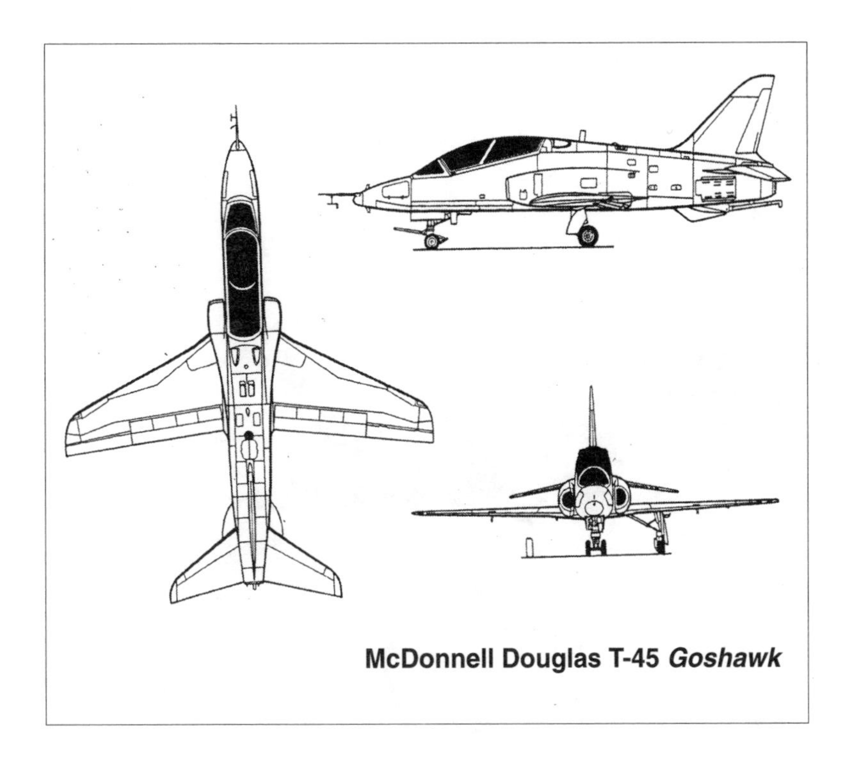

McDonnell Douglas T-45 *Goshawk*

03
03

03
03

Section 7
Fire Fighters

Current Fire Fighters shown in this section are all latter-day applications of original "warbirds." No Fire Fighter aircraft currently seen over southern California were built from the ground up to specifically fulfill the fire fighter role.

Douglas DC-4/C-54/R5D *Skymaster*

Service: All U.S. military branches

Things to look for:

A. Wing mounted low on fuselage.
B. Four radial engines mounted on wings.
C. Wing passes through the middle of the engine nacelles.
D. Oval windows (DC -6/C-118/R6D have square windows).
E. Rounded wing tips and tail (DC-6/C-118/R6D have squared-off wings and tail).

Description:

Designed for commercial transport use, the Douglas DC-4 did not see commercial service until after World War II. During the war it served as the Army Air Corps C-54, and the Navy R5D, with a maximum payload of 28,000 pounds or 49 troops.

The Douglas C-54, a cargo version of the commercial Douglas DC-4, made its first flight on 14 February 1942. More than 1160 C-54 *Skymasters* were built. During World War II the C-54 made 79,642 ocean crossings; only three were lost. C-54s were also used by the Navy, USCG, and USMC under the designator R5D. The C-54/R5D was in use by the U.S. military until the mid-1960s.

Today the California Department of Forestry uses C-54/R5D/DC-5s as fire bombers, dropping fire retardant on brush fires. The DC-6 is also used in this role. The DC-6/C-118/R6D fuselage is 81 inches longer than the DC-4.

Specifications:

Length	93 feet, 10 inches
Wing Span	117 feet, 6 inches
Height	27 feet, 6 inches
Crew	4
Maximum Speed	265 mph
Service Ceiling	22,000 feet
Empty Weight	37,000 pounds
Maximum Weight	62,000 pounds

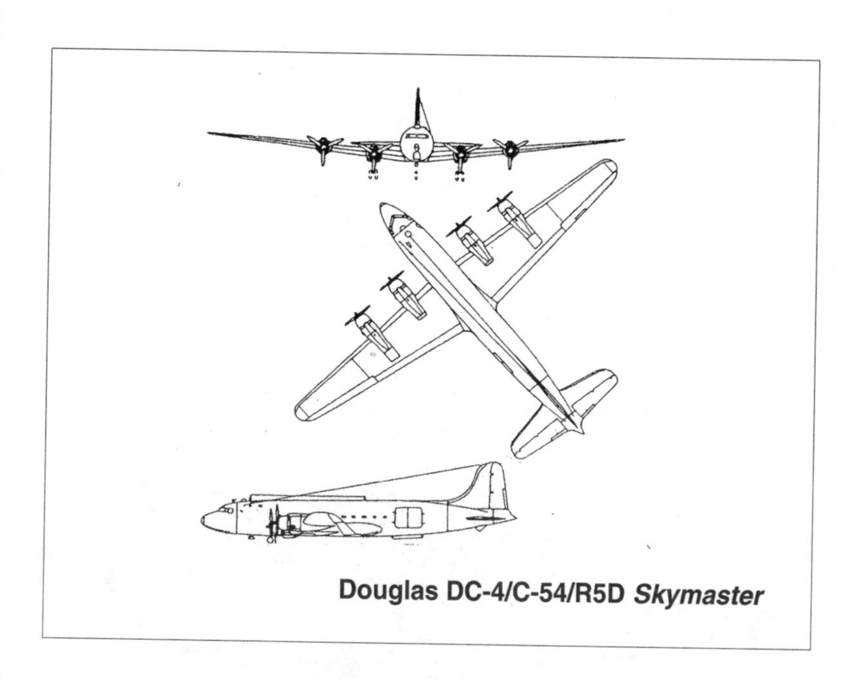

Douglas DC-4/C-54/R5D *Skymaster*

North American OV-10 *Bronco*

Service: U.S. Air Force, Navy, Marine Corps

Things to look for:

A. Twin tail booms.
B. Horizontal tail surfaces located at top of tail.
C. Wings mounted high on fuselage.
D. Cockpit/cargo area located in pod between engine/tail booms.
E. Cockpit canopy is large and bulbous.
F. Sits in a tail-high position.
G. Sponsons located on each side, bottom of fuselage.

Description:

This twin-turboprop aircraft can achieve short takeoffs and landings. Faster than helicopters and slower and more maneuverable than jets, its mission capabilities included observation, forward air control, helicopter escort, armed reconnaissance, gunfire spotting, utility, and limited ground attack.

The OV-10 *Bronco* first flew on 16 July 1965. Production aircraft entered service with the USAF in 1967 and were used in Vietnam in 1968. The U.S. Navy and USMC received 117 OV-10 *Broncos* between 1965 and 1977. The OV-10 can carry 3200 pounds of cargo, five troops, or two litter patients. Desert Storm was the last deployment of the OV-10 where two *Broncos* were lost in combat. In March 1994, Marine Observation Squadron 4 (VMO-4) stood down, passing the *Bronco* to the California Division of Forestry (CDF) and Bureau of Land Management (BLM) where the OV-10's excellent maneuverability and large canopy are advantageous in fighting fires.

Specifications:

Length	41 feet, 7 inches
Wing Span	44 feet
Height	15 feet, 1 inch
Crew	2
Maximum Speed	281
Service Ceiling	24,000 feet
Empty Weight	9893 pounds
Maximum Weight	14,444 pounds

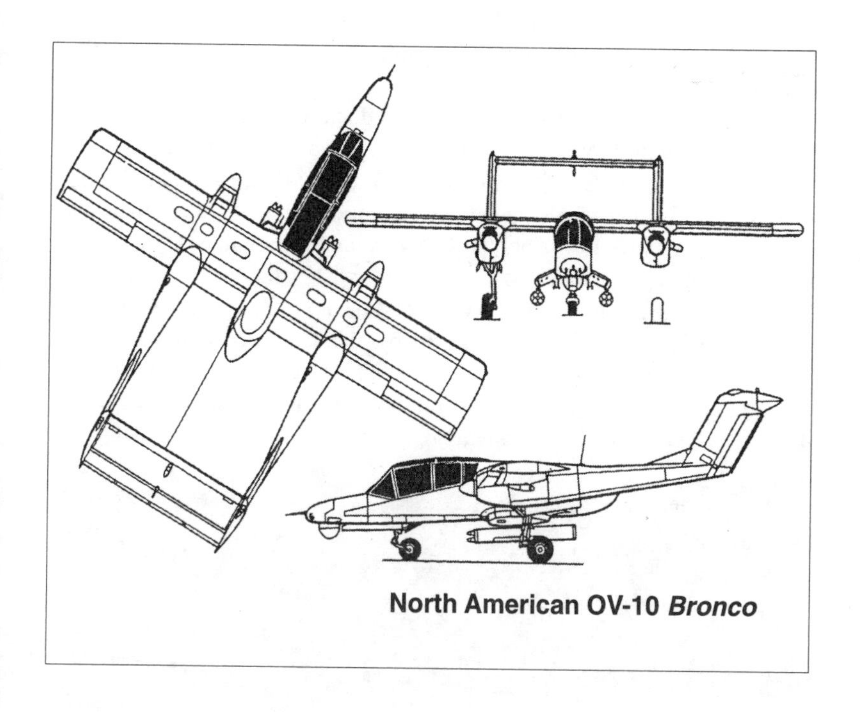

North American OV-10 *Bronco*

Lockheed P2V *Neptune*

Service: U.S. Navy

Things to look for:

A. Pods on the end of each wing.
B. Propeller engines located close to fuselage; two jet engines located in pods on each wing.
C. Tail extends past rear vertical stabilizer.
D. Wings mid-mounted in fuselage.

Description:

The P2V *Neptune* was designed as a long range, land-based patrol bomber for the U.S. Navy but did not see service until after World War II. Once in service, its primary mission became anti-submarine warfare, with a secondary mission of aerial mining.

The P2V first flew on 17 May 1945 and started deliveries to the U.S. Navy in July 1946. More than 1000 P2V *Neptunes* were delivered in seven different major variants before production ended in April 1962. A P2V set a world record for non-stop, non-refueled flight on 29 September 1946. The P2V *Truculent Turtle* flew from Perth, Australia, to Columbus, Ohio, a total of 11,235 miles, and was in the air 55 hours and 17 minutes. The *Truculent Turtle* is on display at the Naval Aviation Museum in Pensacola, Florida.

Today P2V *Neptunes* are used as fire bombers and are stationed all over southern California, ready to fight forest fires. Engines on many (in their civilian roles) have been converted to turboprops, adding many more years of service life to this versatile airframe.

Specifications:

Length	81 feet, 7 inches
Wing Span	103 feet, 10 inches
Height	28 feet, 1 inch
Crew	9
Maximum Speed	323 mph
Service Ceiling	23,200 feet
Empty Weight	41,754 pounds
Maximum Weight	76,152 pounds
Engines	2 Wright R3350 radial engines, 3750 hp each; on some models: 2 Westinghouse J34 turbojet engines rated at 3250 pounds static thrust

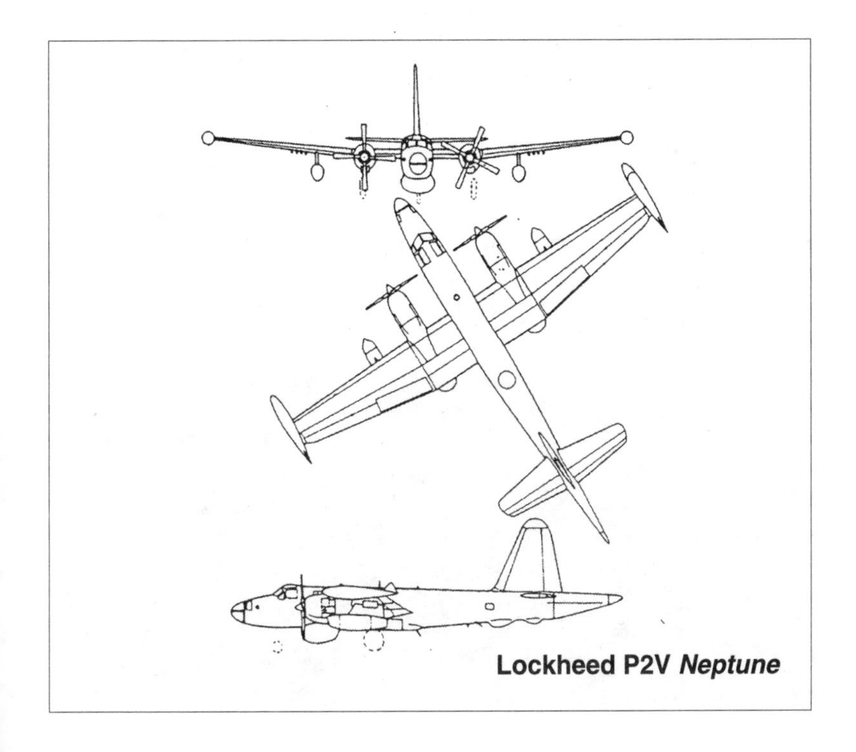

Lockheed P2V *Neptune*

Grumman S-2F *Tracker*

Service: U.S. Navy

Things to look for:

A. Large tail.
B. Shoulder-mounted wings.
C. Twin engines, either radial or turboprop.
D. Horizontal stabilizers on tail located higher than cockpit canopy.
E. Engines and main gear located in "pods' on wings.
F. Engine nacelles extend past the trailing edge of wing.
G. Bulbous cockpit.

Description:

The Grumman S-2F *Tracker* was designed as a hunter/killer aircraft for anti-submarine warfare operations. It became one of the most versatile aircraft in the Navy inventory of its day and was used in carrier on-board delivery, anti-submarine bomber, radar platform, and many other roles. Its folding wings allowed for storage below deck.

The S-2F *Tracker* first flew on 4 December 1952 and more than 1000 were produced. In February 1954 the S-2F entered service with VS-26. The S-2F served as the Navy's submarine search and attack aircraft. The Carrier Onboard Delivery version is known as the C-1 *Trader*. The *Trader* is an S-2F stripped of all of its electronic gear so that it can carry cargo to the carrier from shore. Over 100 S-2F *Trackers* were exported to friendly nations such as Brazil, Argentina, Taiwan, and Thailand. Several S-2Fs have been transferred to the California Division of Forestry. Currently the radial engines in many of these aircraft (in their civilian roles) are being replaced by turboprops.

Specifications:

Length	43 feet, 6 inches
Wing Span	72 feet, 7 inches
Height	16 feet, 7 inches
Crew	6
Maximum Speed	265 mph
Service Ceiling	21,000 feet
Empty Weight	18,750 pounds
Maximum Weight	29,150 pounds
Engines	2 Wright R-1820 radial engines, 1525 hp

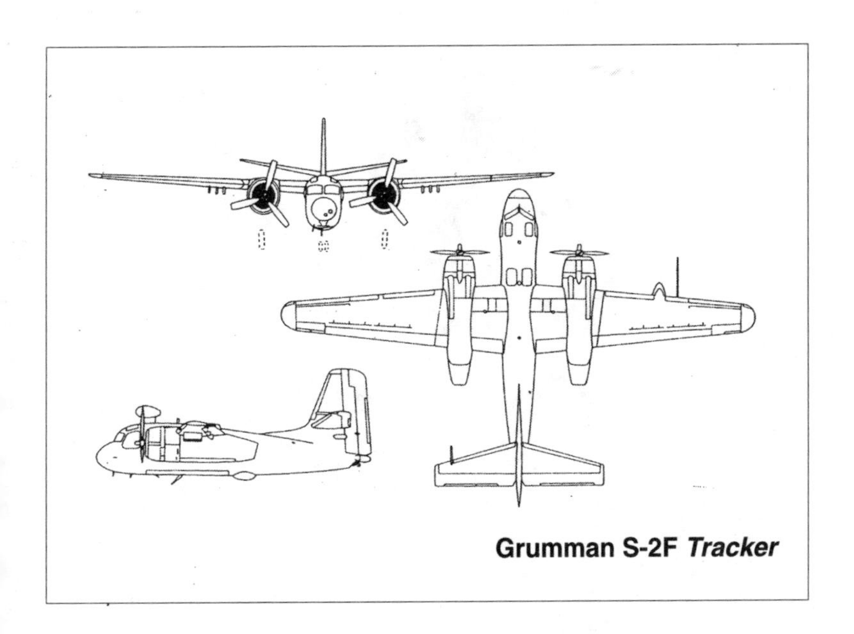

Grumman S-2F *Tracker*

Section 8
Helicopters

Bell AH-1 *Cobra*

Service: U.S. Marine Corps; U.S. Army

Things to look for:

A. Thin fuselage.
B. Stub wings mounted middle of fuselage.
C. Two-bladed main rotor.
D. Gun under nose.
E. Marine version has rounded canopies.
F. Army version has squared canopies.

Description:

The AH-1 *Cobra* is a marginal-weather Marine Corps attack helicopter that provides enroute escort by day or night for Marine assault helicopters and their embarked forces.

The *Cobra* can trace its roots to the UH-1 *Iroquois* or *Huey*. The *Cobra* was a privately funded venture and used the rotors, engine, and transmission from the *Huey*. The *Cobra* first flew on 7 September 1965 and deployed to Vietnam during August 1967. The *Cobra* has had many upgrades and modifications to the basic airframe. Several Marine units flew the AH-1 during Operation Desert Shield/Desert Storm. The Marine Corps has plans to upgrade their 180 AH-1Ws to AH-1Z standards. The AH-1Z will have four-bladed main and tail rotors.

Specifications:

Length—Fuselage	45 feet, 6 inches
Main Rotor Diameter	48 feet
Tail Rotor Diameter	9 feet, 9 inches
Wing Span	10 feet, 7 inches
Height	14 feet, 2 inches
Crew	2
Maximum Speed	175 mph
Service Ceiling	12,000-plus feet
Hover Ceiling	1400 feet in ground effect; 3000 feet out of ground effect
Empty Weight	10,200 pounds
Maximum Take-off Weight	14,750 pounds

Armament: The helicopter carries one 20 mm turreted Gatling cannon. Four external wing stations can carry 2.75-inch and 5.0-inch rockets and a wide variety of precision-guided missiles, including TOW/*Hellfire* (point target/anti-armor) missiles, *Sidewinder* air-to-air missiles, and *Sidearm* anti-radiation missiles.

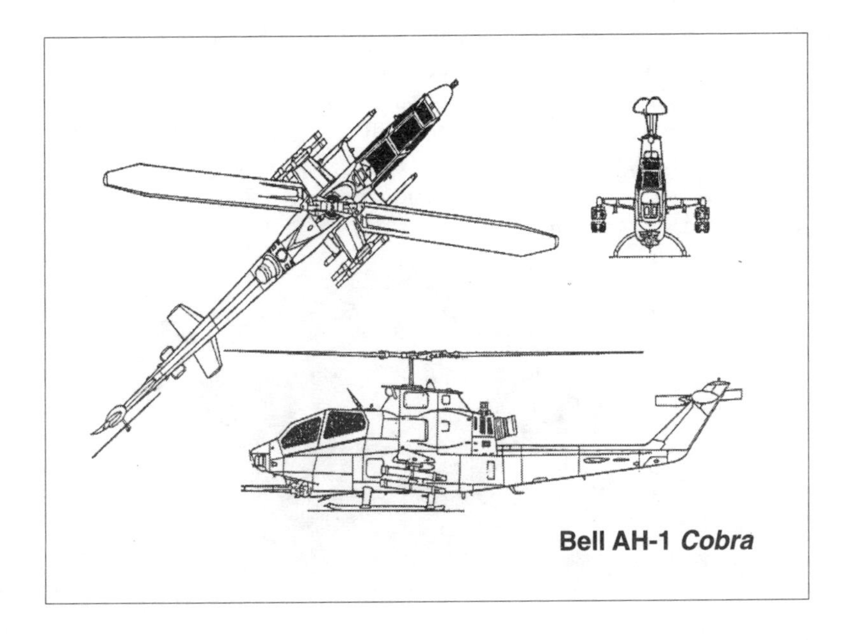

Bell AH-1 *Cobra*

Boeing Vertol CH-46 *Seaknight*

Service: U.S. Marine Corps; U.S. Navy

Things to look for:

A. Two rotors located on top of fuselage.
B. Tricycle landing gear (CH-47 has four main landing gear).
C. Short pods located on each side of fuselage (CH-47 pods are almost full length of fuselage).
D. Engines located on top of rear fuselage.
E. Cargo ramp located at rear of fuselage.
F. Lower rear surface of fuselage slants upward.

Description:

This medium-lift helicopter is used by the Navy for shipboard delivery of cargo and personnel. It is used by the Marine Corps to provide all-weather, day-or-night assault transport of combat troops, supplies, and equipment. Troop assault is the primary function and the movement of supplies and equipment is secondary. Additional tasks may be assigned, such as combat support, search and rescue, support for forward refueling and re-arming points, aeromedic evacuation of casualties from the field, and recovery of aircraft and personnel.

The CH-46 *Seaknight* first flew on 22 April 1958. The Army had looked at the CH-46 but elected to acquire the *Seaknight's* bigger brother, the Boeing CH-47 *Chinook*. The first CH-46's *Seaknights* entered service with the Marine Corps in June 1964. The *Seaknight* will eventually be replaced by the MV-22 *Osprey*. The CH-46 is capable of carrying 24 passengers, 17 combat equipped troops, or 6000 pounds of cargo. Versions of the CH-46 are operated by Sweden, Japan, and Canada.

Specifications:

Overall Length	84 feet, 4 inches (rotors opened); 45 feet 8 inches (rotors folded)
Length—Fuselage	45 feet, 8 inches
Rotor Diameter	51 feet
Height	16 feet, 8 inches
Crew	3
Maximum Speed	166 mph
Service Ceiling	14,000 feet
Empty Weight	15,537 pounds
Maximum Weight	24,300 pounds
Armament	None

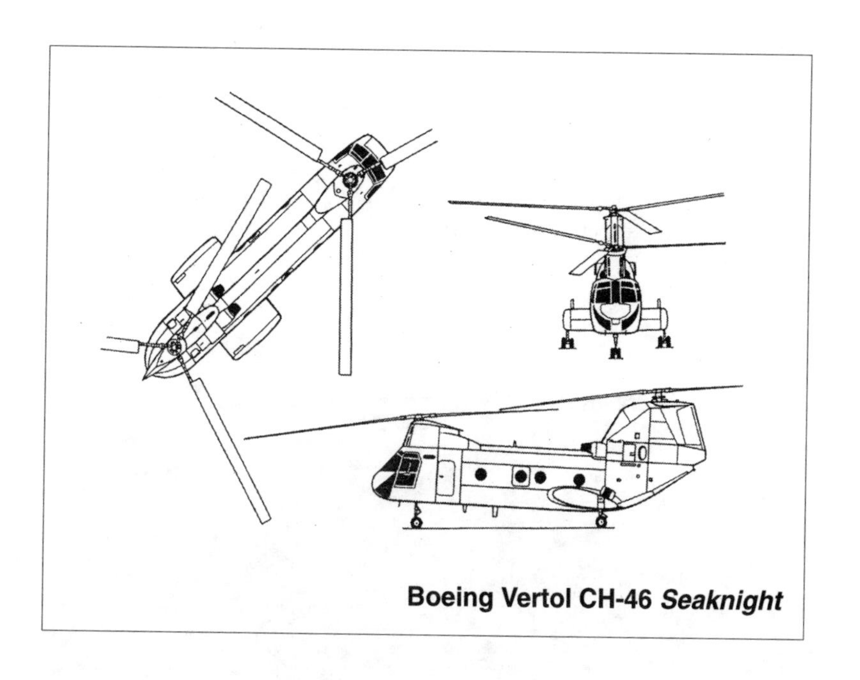

Boeing Vertol CH-46 *Seaknight*

Sikorsky CH-53E *Super Stallion*

Service: U.S. Navy; U.S. Marine Corps

Things to look for:

A. Tail canted to the left.
B. Third engine located behind rotor mast, left side.
C. Sponsons located on each side.
D. Seven blades on main rotor.
E. Refueling probe protubes from front of helicopter.
F. Wide tail.

Description:

This heavy-lift helicopter is designed to transport personnel, supplies, and equipment in support of amphibious and shore operations. It is slated for replacement by the MV-22 *Osprey*.

The CH-53E first flew on 1 March 1974. The CH-53E *Super Stallion* differs from earlier models by having a third engine placed behind the rotor mast, and by having another rotor blade added for a total of seven rotor blades. The tail of the CH-53E is canted 20 degrees to the left. The *Super Stallion* is the heaviest lift helicopter in the western world. It is able to carry 55 troops or 32,000 pounds of cargo. The Navy and Marine Corps are the only operators of the CH-53E. The U.S. Navy and the Japanese Maritime Self Defense Force operate the mine-clearing version of the CH-53E, known as the MH-53E.

Specifications:

Length	99 feet, 0.5 inch (rotors turning); 60 feet, 6 inches (tail and rotors folded)
Length—Fuselage	73 feet, 4 inches
Height	29 feet, 5 inches
Crew	3
Maximum Speed	196 mph
Service Ceiling	21,000 feet
Empty Weight	24,606 pounds
Maximum Weight	42,000 pounds

Armament: Two 0.5-inch machine guns

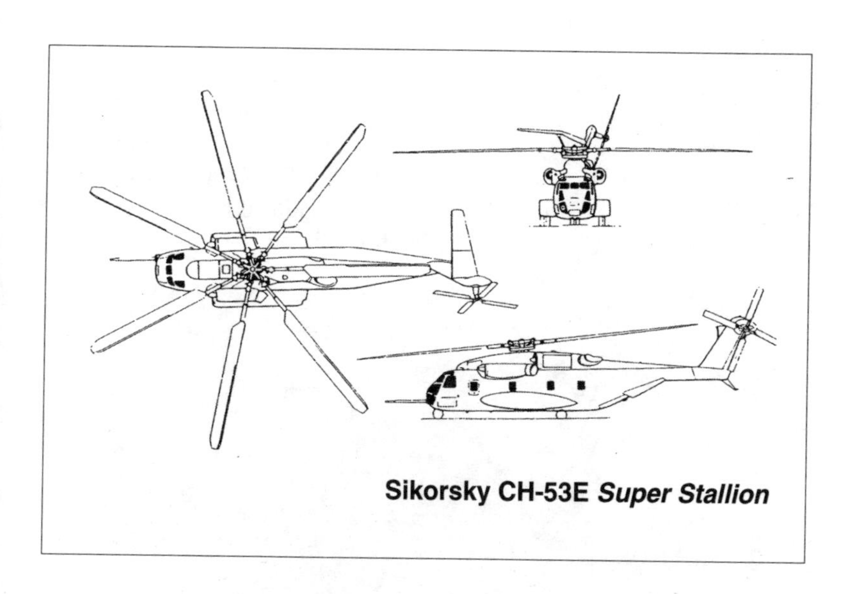

Sikorsky CH-53E *Super Stallion*

Sikorsky CH-54 *Tarhe*, aka *Sky Crane*

Service: **U.S. Army**

Things to look for:

A. Extremely thin fuselage—looks like a spine.
B. Cockpit located in pod dropping below fuselage.
C. Main landing gear located midway of fuselage, support strut slopes down at 45-degree angle.
D. Engines located on top of fuselage above main landing gear.

Description:

The CH-54 was named for Chief Tarhe, a distinguished Wyandot chief. He was nicknamed "Crane" (a tall fowl), which may have had some bearing on the naming of this aircraft. The CH-54 was designed for heavy internal loads or external lift of heavy bulk loads. It was able to carry interchangeable pods for troop or missile transport, mine laying, cargo, and field hospital operations. A hoist was provided to allow pickups and deliveries without landing. It had a secondary rear-facing pilot's station to provide a clear view of the cargo.

The CH-54 first flew on 9 May 1962. Deployed to Vietnam in 1965, the CH-54 is credited with retrieving more than 350 downed aircraft. It was also used as a bomber carrying the 10,000-pound *Daisy Cutter* bomb to clear landing areas. In the late 1960s and early 1970s the CH-54 was replaced by the CH-47 *Chinook* and was transferred to the National Guard and Army Reserve. Today many CH-54s are used in fire fighting.

Specifications:

Length	88 feet, 6 inches (rotors turning)
Length—Fuselage	70 feet, 3 inches
Rotor Diameter	72 feet
Height	18 feet, 7 inches (top of rotor mast)
Crew	3–4
Maximum Speed	126 mph
Service Ceiling	9000 feet
Empty Weight	19,980 pounds
Maximum Weight	47,000 pounds
Armament	None

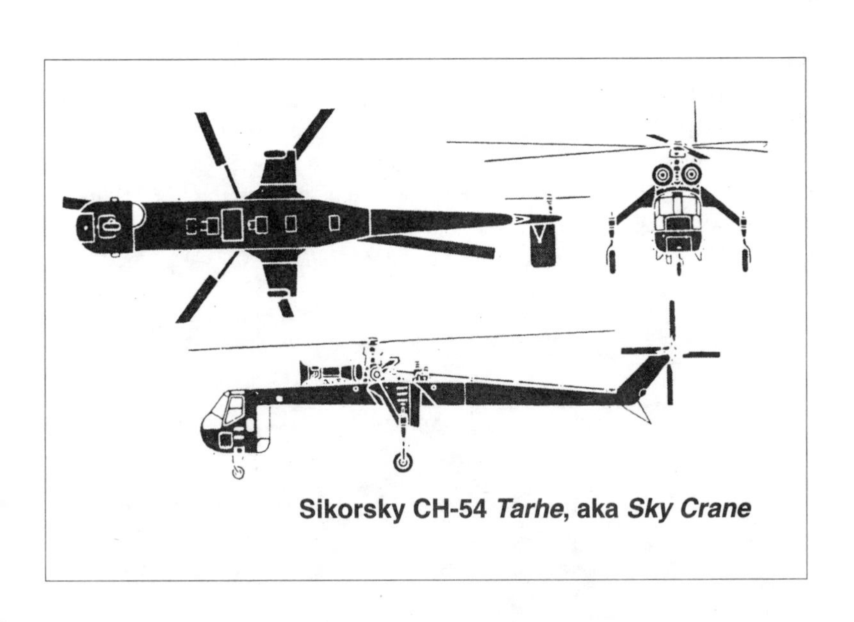

Sikorsky CH-54 *Tarhe*, aka *Sky Crane*

Eurocopter (Aerospatiale) HH-65 *Dolphin*

Service: U.S. Coast Guard

Things to look for:

A. Shrouded tail rotor.
B. Tail extends above the rotor blades.
C. Tail has three vertical stabilizers.
D. Wheels fully retractable.
E. Hoist located over right (starboard) side of cabin.
F. Engines and rotors enclosed in one large component over cabin.

Description:

The HH-65 *Dolphin*, a medium-range recovery helicopter, is used for SRR (Short Range Recovery), SAR (Search and Rescue), narcotics interdiction, law enforcement, environmental protection, and airlift support of polar operations. The HH-65 *Dolphin* is capable of flying automatic search patterns, thereby allowing the pilot and co-pilot to aid in visual search.

The HH-65 first flew on 23 July 1980 in France. The aircraft was then shipped to Grand Prairie, Texas, for fitting of U.S. equipment to satisfy political requirements. The first model was delivered to the U.S. Coast Guard on 1 February 1987. The *Dolphin* is fitted with a composite rotor head, fiberglass rotor blades, and composite material structural assemblies, and thus has earned the name *Plastic Puppy*. More than 90 *Dolphins* were purchased by the U.S. Coast Guard.

Photo by Ray Rivard

Specifications:

Overall Length	45 feet, 6 inches
Length—Fuselage	38 feet, 2 inches
Rotor Diameter	39 feet, 2 inches
Height	13 feet
Crew	3
Maximum Speed	165 kph
Empty Weight	6092 pounds
Maximum Weight	9200 pounds
Armament	None

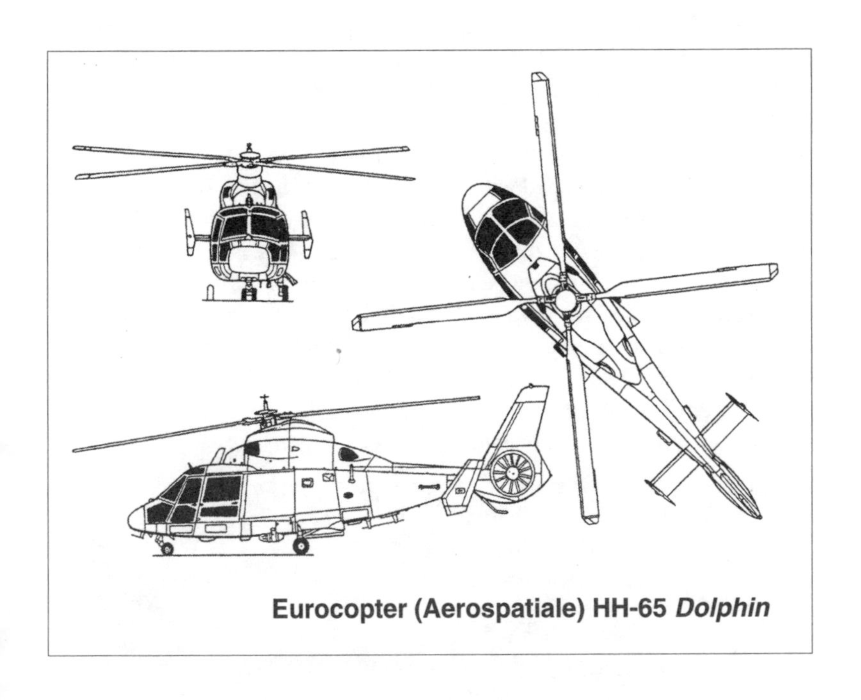

Eurocopter (Aerospatiale) HH-65 *Dolphin*

Kaman SH-2 *Seasprite*

Service: **U.S. Navy**

Things to look for:

A. Engines located in pods over cockpit.
B. Very distinct hump over cockpit.
C. Round radar radome located under nose.
D. Wheel wells located under cabin door.
E. Horizontal tail surfaces have support struts.

Description:

The SH-2 *Seasprite* was designed for a variety of missions. Primary missions include anti-submarine and anti-surface warfare, anti-ship missile defense, and anti-ship surveillance and targeting. Secondary missions include medical evacuation, search and rescue, personnel and cargo transfer, and vertical replenishment (VERTREP), as well as small boat interdiction, amphibious assault air support, gunfire spotting, mine detection, and battle damage assessment.

The SH-2 *Seasprite* entered service in December 1962. The original SH-2 *Seasprite* was equipped with a single engine. The SH-2C was the first model equipped with two engines. The SH-2 *Seasprite* was the first Kaman-built helicopter equipped with the standard main rotor/tail rotor setup instead of intermeshing side-by-side rotors. The *Seasprite* is also equipped with retractable wheels and a flotation hull that allow it to operate directly from the water. The *Seasprite* still serves aboard Navy ships that are not able to handle the SH-60 *Seahawk*. During 1989 the first SH-2G *Super Seasprite* took to the air. The *Super Seasprite* is equipped with General Electric T-700 turboshaft engines and will eventually replace the *Seasprite*.

Specifications:

Overall Length	52 feet, 9 inches
Length—Fuselage	40 feet, 6 inches
Rotor Diameter	44 feet, 4 inches
Height	15 feet, 0.5 inch
Crew	3
Maximum Speed	165 mph
Empty Weight	7600 pounds
Maximum Weight	13,500 pounds

Armament: Two Mk-46 or Mk-50 torpedoes

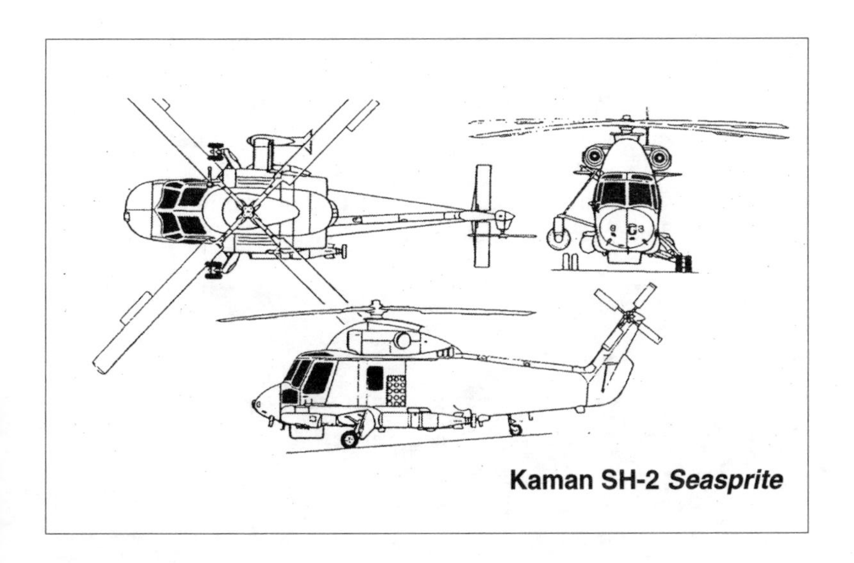

Kaman SH-2 *Seasprite*

Sikorsky SH-3 *Sea King*

Service: U.S. Navy, U.S. Marine Corps

Things to look for:

A. Wheels located in pods on side of fuselage.
B. Bracing from fuselage to wheel pods.
C. Engines located on top of cockpit.
D. Short straight tail boom.
E. Horizontal tail surfaces, even with tail rotor.
F. Main rotor has five blades.

Description:

This twin-engine, all-weather helicopter is used by the Navy Reserves to detect, classify, track, and destroy enemy submarines. The UH-3H model is utility-configured for logistical support and search and rescue missions.

The SH-3 *Sea King* first flew on 11 March 1959. Originally designated as the HSS, it was redesignated the SH-3 in 1962. The *Sea King* was first delivered to VHS-3 in Norfolk, Virginia, and VHS-10 in San Diego, California, during September 1961. The *Sea King* was designed to replace the anti-submarine warfare (ASW) hunter-killer teams of the 1950s. It allowed the sensor equipment and weapons to be carried in one aircraft instead of two. HMX-1 is the only USMC squadron to operate the *Sea King*. *Marine One*, as it is known, is the aircraft used for the flights of the President. The SH-3 *Sea King* has almost been completely replaced by the SH-60 *Sea Hawk*.

Photo by Ray Rivard

Specifications:

Overall Length	72 feet, 8 inches (rotors turning)
Length—Fuselage	54 feet, 9 inches
Length with Tail Folded	47 feet, 3 inches
Height	16 feet, 10 inches
Crew	4
Maximum Speed	166 mph
Service Ceiling	14,700 feet
Empty Weight	12,350 pounds
Maximum Weight	21,000 pounds

Armament: Two Mk-46 torpedoes

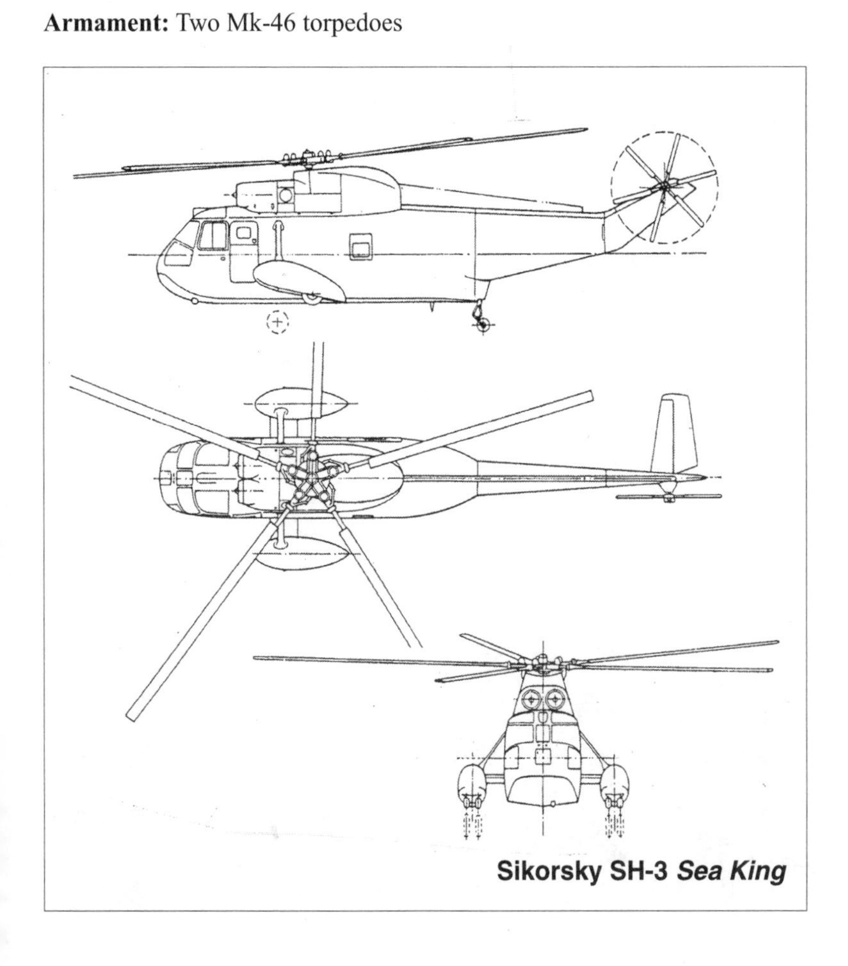

Sikorsky SH-3 *Sea King*

Sikorsky SH-60 *Seahawk*; UH-60 *Blackhawk*; HH-60J *Jayhawk*

Service: U.S. Army, Navy, Air Force, Coast Guard

Things to look for:

A. Wheels instead of skids.
B. Rotors have swept-back tips.
C. Fuselage is short, squat.
D. Large vertical stabilizer located under tail rotor.
E. Tail wheel of Army version located towards end of tail boom.
F. Tail wheel of Navy/Coast Guard version located towards front of tail boom.
G. Tail rotor canted upwards.

Description:

The SH-60 *Seahawk* is a twin-engine helicopter that is used for anti-submarine warfare, search and rescue, drug interdiction, anti-ship warfare, cargo lift, and special operations.

The UH-60 *Blackhawk* entered service with the 101st Airborne Division in June 1979. The Army uses *Blackhawk* in the assault/utility role. It can carry up to 11 troops or 4 litter patients. The navalized version of *Blackhawk* first flew in December 1979 and has much in common with the Army version. The Navy version is equipped with folding tail boom and main rotor and has different landing gear.

The *Jayhawk*, the Coast Guard version of the SH-60 *Seahawk*, is equipped with dedicated search equipment instead of Navy combat gear.

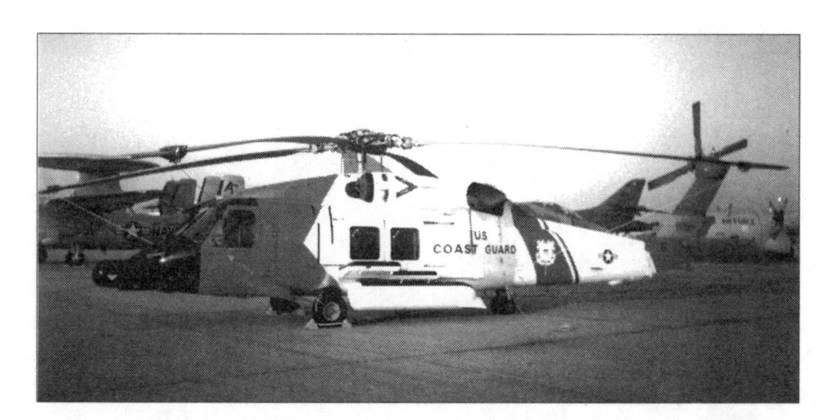

Specifications:

Length	64 feet, 10 inches; tail folded: 40 feet, 11 inches
Rotor Diameter	53 feet, 8 inches
Height	12 feet, 5 inches
Crew	3
Maximum Speed	184 mph
Empty Weight	13,648 pounds
Maximum Weight	21,884 pounds

Armament: The *Seahawk* usually carries two 7.62 mm machine guns mounted in the windows. It can also be equipped with *Hellfire* air-to-surface or *Penguin* anti-ship missiles, three Mk-46 or Mk-50 torpedoes, or additional 0.5-inch machine guns mounted in the doors.

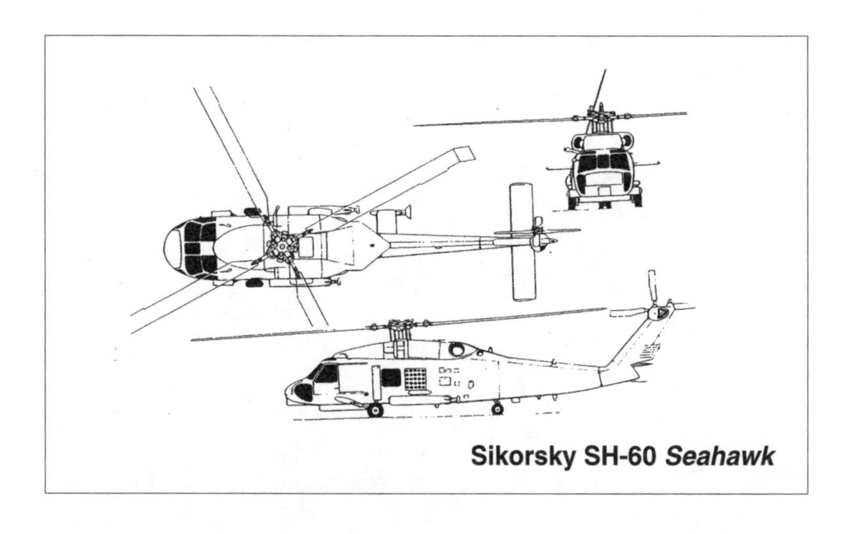

Sikorsky SH-60 *Seahawk*

Bell UH-1 *Iroquois*, aka *Huey*

Service: U.S. Army, Navy, Marine Corps

Things to look for:

A. Two-bladed rotor blade.
B. Swept-back tail.
C. Horizontal tail located on tail boom.
D. Engine(s) located on top of fuselage.

Description:

This twin-piloted, twin-engine helicopter is used in command and control, resupply, casualty evacuation, and troop transport. It provides utility combat helicopter support to the landing force commander during ship-to-shore movement and in subsequent operations ashore.

The UH-1 *Iroquois* made its first flight in October 1956 and was originally designated the HU-1. During 1959 it was redesignated the UH-1. The *Iroquois*, or *Huey*, first entered Army service in 1958 and the Army has procured more than 9000 *Hueys* in the last 30-plus years. The USMC received its first *Huey* in 1964. The UH-1N entered Marine Corps service in 1971. The Corps plans upgrades to keep the UH-1 flying until 2020. The Army has been replacing their *Hueys* with the UH-60 *Blackhawk* in front-line units.

Specifications:

Overall Length	57 feet, 1 inch
Length—Fuselage	44 feet, 6 inches
Main Rotor Diameter	48 feet
Height	14 feet, 7 inches
Crew	3–4
Maximum Speed	127 mph
Service Ceiling	18,500 feet
Empty Weight	4717 pounds
Maximum Weight	9500 pounds

Armament: The *Iroquois* can carry an M-240 7.62 mm machine gun, or a GAU-16 50 caliber machine gun, or a GAU-17 7.62 mm automatic gun. All three weapons' systems are crew-served, and the GAU-2B/A can also be controlled by the pilot in the fixed forward firing mode. The helicopter can also carry two 7-shot, or 19-shot, 2.75-inch rocket pods.

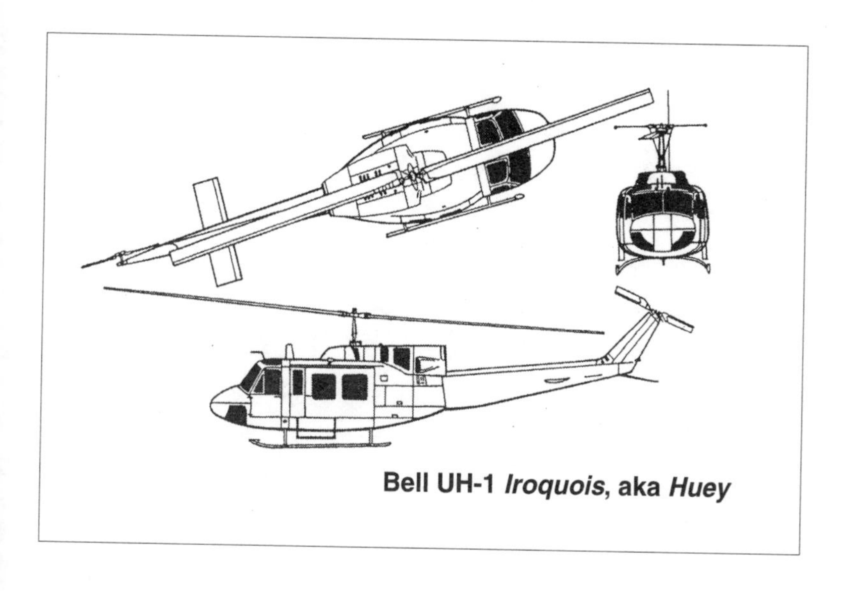

Bell UH-1 *Iroquois*, aka *Huey*

Section 9
Restored Aircraft

North American AT-6/SNJ *Texan*

Service: **U.S. Air Force; U.S. Marine Corps; U.S. Navy; U.S. Army Air Corps**

Things to look for:

A. Wings mounted low on fuselage.
B. Long cockpit canopy.
C. Tandem seat cockpit.
D. Triangular-shaped tail.
E. Leading edge of wings are swept back.

Description:

The AT-6 *Texan* is a two-place advanced trainer that was designed as a transition trainer between basic trainers and first-line tactical aircraft. It has filled many different roles for which it was not originally designed, such as fighter, interceptor, fighter-bomber, forward air controller (FAC), and counter-insurgency platform (COIN). It has seen combat in World War II, Korea, and Vietnam. Most of the Allied pilots of World War II were trained in this airplane. It has also been used by many foreign government air forces.

The T-6/SNJ *Texan* first flew 1 April 1935. It has been manufactured by Australia, Sweden, Canada, and other countries under license, with over 15,000 being built. Several AT-6/SNJ *Texans*, in flying condition and static display, are located in the San Diego area. It has even appeared in movies. It was replaced by the T-28 *Trojan* and later the T-34 *Mentor* as a trainer for the U.S. Navy and Marine Corps.

Specifications:

Length	29 feet, 6 inches
Wing Span	42 feet, 5 inches
Height	11 feet, 9 inches
Crew	2
Maximum Speed	205 mph
Service Ceiling	21,500 feet
Empty Weight	4158 pounds
Maximum Weight	5300 pounds
Engine	Pratt & Whitney R1340, 9-cylinder radial engine
Armament	None

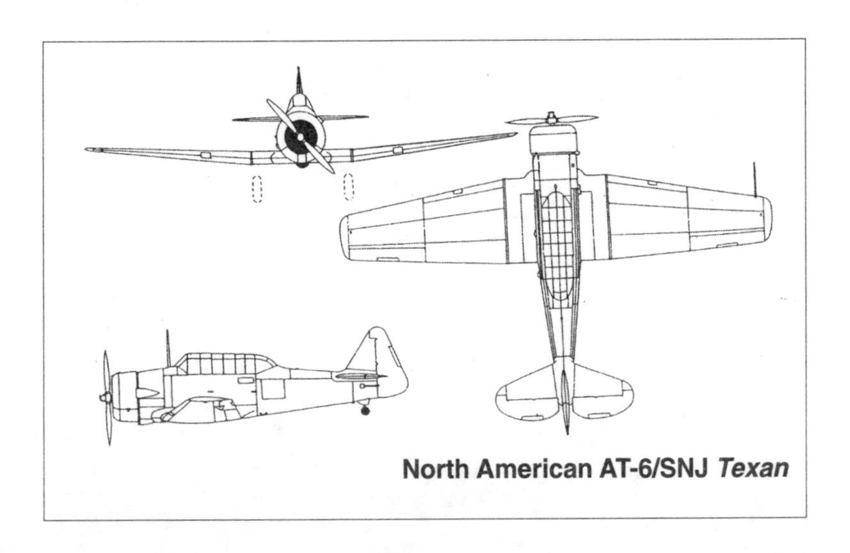

North American AT-6/SNJ *Texan*

Boeing B-17 *Flying Fortress*

Service: U.S. Air Force, U.S. Navy

Things to look for:

A. Four radial engines located on the wing.
B. Wing mounted low on fuselage.
C. Main landing gear did not fully retract into engine nacelles.
D. Leading edge of wing had a slight taper to it.
E. Plexiglass nose.
F. Waist gunner positions located on each side of fuselage, between wing and tail.

Description:

The B-17 was one of the United States' two standard heavy bombers until the introduction of the B-29 *Superfortress*. It served in every WW II combat zone but is best known for daylight strategic bombing of German industrial targets. It was also used for search and rescue (SAR) and as an anti-submarine platform. The B-17 also served as a VIP carrier in Korea, and served early in the Vietnam War, inserting agents into North Vietnam.

The B-17 *Flying Fortress* first flew 28 July 1935. There were six basic models of the B-17 and a total of 12,716 were built. The Navy operated the B-17 as the PB-1. It was also used as a fire bomber after the war and was retired from that role in the early 1970s.

Specifications:

Length	74 feet, 4 inches
Wing Span	103 feet, 9.375 inches
Height	19 feet, 2.5 inches
Crew	10
Maximum Speed	302 mph
Service Ceiling	36,400 feet
Empty Weight	36,134 pounds
Maximum Weight	40,260 pounds
Engines	4 Wright Cyclone R1820 radial engines, 1000 hp each

Armament: 11 to 13 machine guns; 20,000-pound bomb load (B-17G)

Boeing B-17 *Flying Fortress*

North American B-25/PBJ *Mitchell*

Service: U.S. Air Force; U.S. Marine Corps; U.S. Navy; U.S. Army Air Corps

Things to look for:

A. Rectangular twin tails.
B. Twin engine.
C. Engine nacelle extends beyond trailing edge of wing.
D. Wings mounted midway on fuselage.
E. Inverted "gull" wings.
F. Main landing gear located in engine nacelles.

Description:

A highly versatile twin-engine bomber, the B-25 was used by the U.S. Air Force during World War II for high- and low-level bombing, strafing, photo reconnaissance, and submarine patrol. It was even used as a fighter.

The B-25 *Mitchell* first flew on 19 August 1940. There were six major models of the B-25 and more than 11,000 were built. The B-25's most famous operation was the Doolittle Raid of 18 April 1942. Sixteen B-25Bs were launched from the aircraft carrier USS *Hornet* (CV-8) and carried out the first air attack on mainland Japan.

The Navy version of the B-25, designated PBJ, was used for patrol and ground attack. The major differences in the B-25 and PBJ were in the avionics and radio packages. After World War II, B-25s continued to serve until the mid-1950s as trainers, VIP transports, and squadron hacks. Today there are several B-25s flying and on static display in the San Diego area.

Specifications:

Length	52 feet, 11 inches
Wing Span	67 feet, 7 inches
Height	15 feet, 9 inches
Crew	5
Maximum Speed	275 mph
Service Ceiling	23,800 feet
Empty Weight	16,000 pounds
Maximum Weight	35,000 pounds
Engines	2 Wright R2600 radial engines, 1350 hp each

Armament: The B-25 standard version carries one 0.30-inch machine gun in a flexible mount in the nose, one 0.30-inch machine gun in a flexible dorsal position, one 0.30-inch machine gun in a flexible waist position, and one 0.50-inch machine gun in a flexible tail position. The nominal bomb load is 3000 pounds.

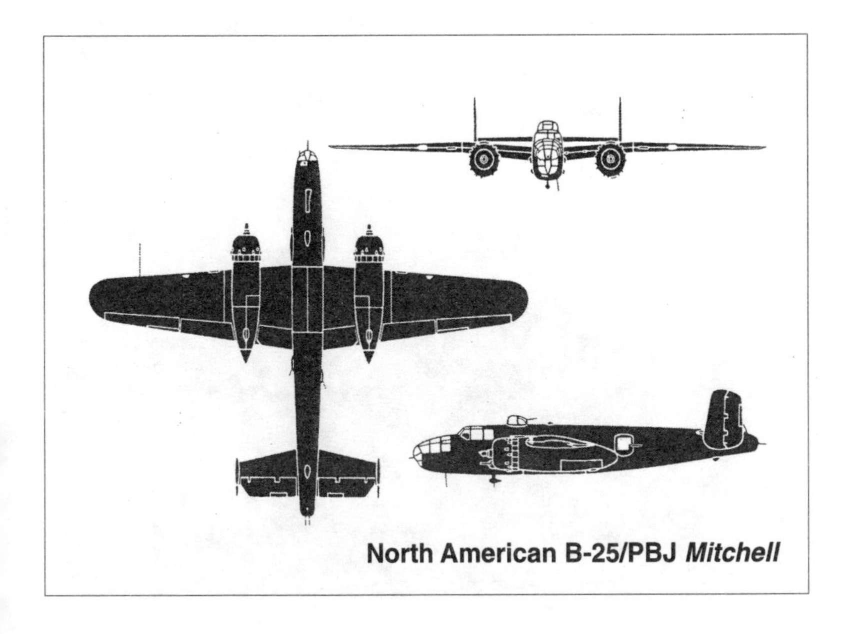

North American B-25/PBJ *Mitchell*

Grumman F-4F/FM-2 *Wildcat*

Service: U.S. Navy; U.S. Marine Corps

Things to look for:

A. Short stubby fuselage.
B. Wing mounted in middle of fuselage.
C. Wheels flat against fuselage under wing when retracted.
D. Air scoop located under right or starboard wing.
E. Landing gear has very narrow track when extended.
F. Wing tips and tail surfaces squared off.

Description:

The Grumman F-4F *Wildcat* was the standard fighter of the U.S. Navy at the start of the Pacific War. A mid-wing monoplane, it operated from aircraft carriers and shore bases, provided air cover for friendly forces, spear-headed strikes on enemy vessels or shore installations, could withstand a great deal of battle damage, and was a very stable gun platform.

The F-4F first flew on 12 February 1939. The F-4F *Wildcat* had its introduction to combat during the battle of Wake Island. During the first Japanese air strike, seven of VMF-211's twelve F-4F *Wildcats* were destroyed. The F-4F *Wildcat* produced more Medal of Honor winners than any other single engine fighter of World War II. About three quarters of the 8000 F-4Fs built were produced by the Eastern Aircraft Division of General Motors and were designated FM-1 and FM-2. Today, Air Group One located at Gillespie field currently has a FM-2 *Wildcat* in flying condition.

Specifications:

Length	28 feet, 10 inches
Wing Span	38 feet
Height	11 feet
Crew	1
Maximum Speed	318 mph
Service Ceiling	34,900 feet
Maximum Weight	7952 pounds

Armament: Six 0.5-inch Browning machine guns in the outer wings; two 250-pound bombs on the underwing racks

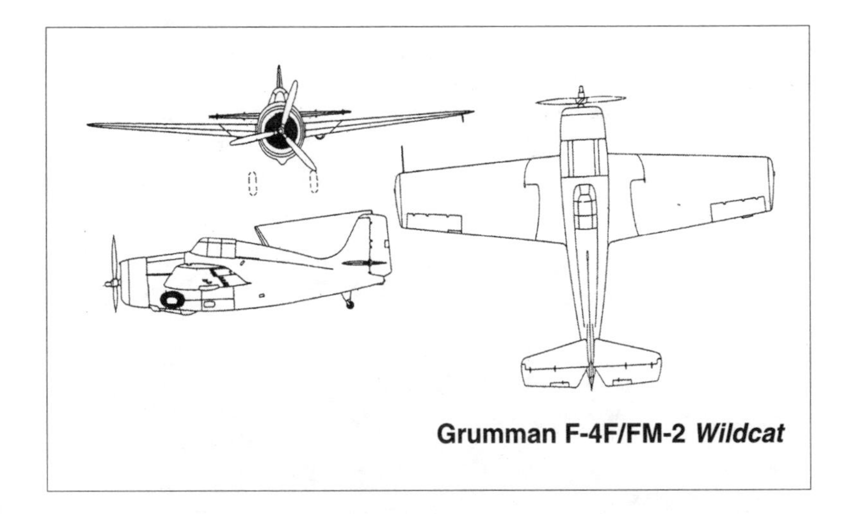

Grumman F-4F/FM-2 *Wildcat*

Chance Vought F-4U *Corsair*

Service: **U.S. Navy; U.S. Marine Corps**

Things to look for:

A. Gull wings.
B. Long nose between cockpit and propeller.
C. Main landing gear located at bottom of curved wings.
D. Round engine cowling.

Description:

Originally designed as a high-altitude fighter, the Chance Vought F-4U *Corsair* was the most powerful of the the World War II Navy fighters. It enhanced the usual fighter missions—providing air cover for friendly forces and spearheading strikes on enemy vessels or shore installations—by becoming a fighter-bomber and night fighter. It was said, "...it could oufight, outclimb, and if need be, outrun any prop driven enemy."

The F-4U *Corsair* first flew on 29 May 1940 and was one of the most popular aircraft from the World War II and Korean War eras. During World War II, Goodyear built the FG-1 *Corsair*, Brewster built the F-3A *Corsair*, and Chance Vought built the F-4U *Corsair*. The *Corsair* was in production from 1942 to 1952, the longest production run time of any American piston-powered fighter. A total of 12,571 *Corsairs* were built and more than 50 survive today. Surviving *Corsairs* can be seen at the Flying Leatherneck Aviation Museum and the Planes of Fame Air Museum.

Specifications:

Length	34 feet, 6 inches
Wing Span	41 feet
Height	14 feet, 9 inches
Crew	1
Maximum Speed	462 mph
Engine	Pratt & Whitney R2800 radial engine, 2000 hp

Armament: Six 0.5-inch wing-mounted machine guns; two 1000-pound bombs

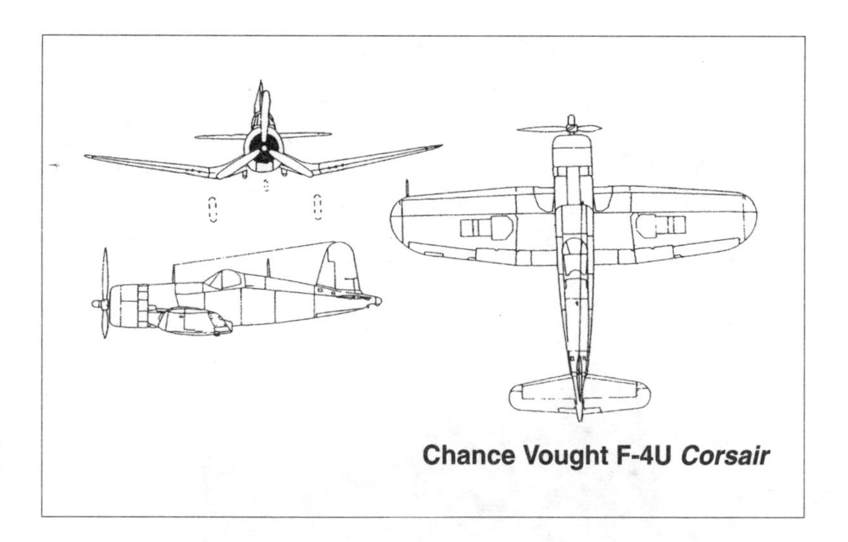

Chance Vought F-4U *Corsair*

Grumman F6F *Hellcat*

Service: U.S. Navy, U.S. Marine Corps

Things to look for:

A. Three-bladed propeller.
B. Fuselage slopes downward from cockpit to tail.
C. Outboard of wings—dihedral (cranked up).
D. Wing mounted just below middle of fuselage.
E. Oil-cooling scoops located in bottom portion of engine cowling.

Description:

The Grumman F6F (Fighter, 6th model, F-Grumman) *Hellcat* was a single-place, carrier-based fighter designed to replace the Grumman F-4F *Wildcat*. More heavily armed, it was speedier, had an excellent rate of climb, and was highly maneuverable.

After entering the fleet, the F6F *Hellcat* made its first kill on 1 September 1943 and soon became the most successful Naval fighter of World War II. Almost 75 percent of the Navy's air-to-air victories were by *Hellcats*. The F6F *Hellcat* produced more aces than any other single engine aircraft—307 aces in all (the P-51 *Mustang* produced 275 aces). The Grumman F6F *Hellcat* ended the war with a 19:1 kill ratio—5155 enemy aircraft destroyed with a loss of only 270 *Hellcats*.

The F6F first flew on 26 June 1942 and deliveries started in 1943 with VF-9 aboard USS *Essex* (CV-9). The Planes of Fame Air Museum in Chino and Palm Springs Air Museum both operate flying *Hellcats*. San Diego Aerospace Museum in Balboa Park has one on display.

Specifications:

Length	33 feet, 7 inches
Wing Span	42 feet, 10 inches
Height	13 feet, 1inch
Crew	1
Maximum Speed	380 mph
Service Ceiling	37,300 feet
Empty Weight	9238 pounds
Maximum Weight	15,413 pounds
Engine	1 Pratt & Whitney R-2800 engine, 2000 hp

Armament: Six 0.5-inch wing-mounted machine guns; two 1000-pound bombs or six 5-inch rockets

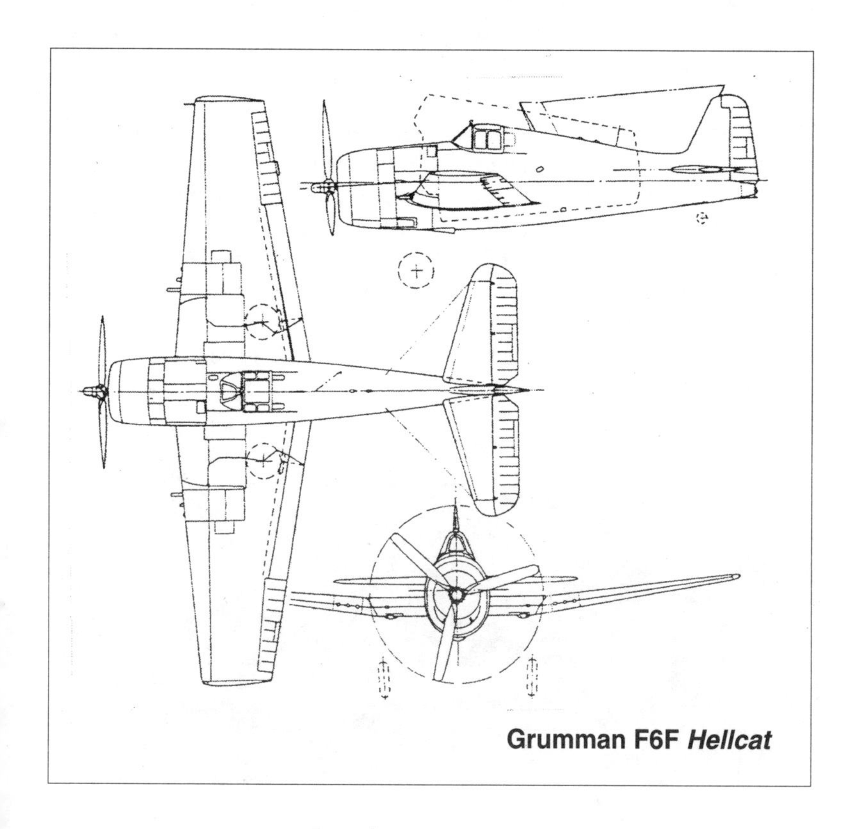

Grumman F6F *Hellcat*

North American F-82 *Twin Mustang*

Service: U.S. Air Force

Things to look for:

A. Two identical P-51 fuselages.
B. Air scoops located under each fuselage.
C. Squared-off wing tips.
D. Horizontal tail surfaces do not extend beyond the vertical tail surfaces.
E. Bubble canopies.
F. Wing mounted low on fuselage.

Description:

This long-range escort fighter was the last propeller-driven production fighter acquired by the USAF. It was designed to alleviate pilot fatigue on long-range bomber escort missions by carrying a co-pilot/navigator.

The F-82 *Twin Mustang* first flew on 16 June 1945. Unlike the P-38 *Lightning*, propellers of the F-82 *Twin Mustang* rotated towards the fuselage (the P-38 propellers rotated away from the fuselage). Using the basic fuselage of the P-51F, the fuselages were lengthened 57 inches and mated together using a constant chord (or width) inner wing. The inner wing carried the aircraft's armament. The port cockpit was equipped with flight controls; the starboard cockpit was occupied by a radar operator.

Arriving too late for World War II, the F-82 was extensively used in the Korean conflict. In fact, an F-82 recorded the first official kill of the Korean War on 27 June 1950, when a *Twin Mustang* of the 68th All Weather Fighter Squadron shot down a Russian YAK-11. An F-82B is being restored at Gillespie Field, El Cajon.

Specifications:

Length	39 feet, 1 inch
Wing Span	51 feet, 3 inches
Height	13 feet, 10 inches
Crew	2
Maximum Speed	482 mph
Service Ceiling	41,600 feet
Empty Weight	13,405 pounds
Maximum Weight	22,000 pounds
Engines	2 Packard Merlin V-1650 in-line engines, 1860 hp each

Armament: Six 0.5-inch machine guns, 25 five-inch rockets, and 4000 pounds of bombs

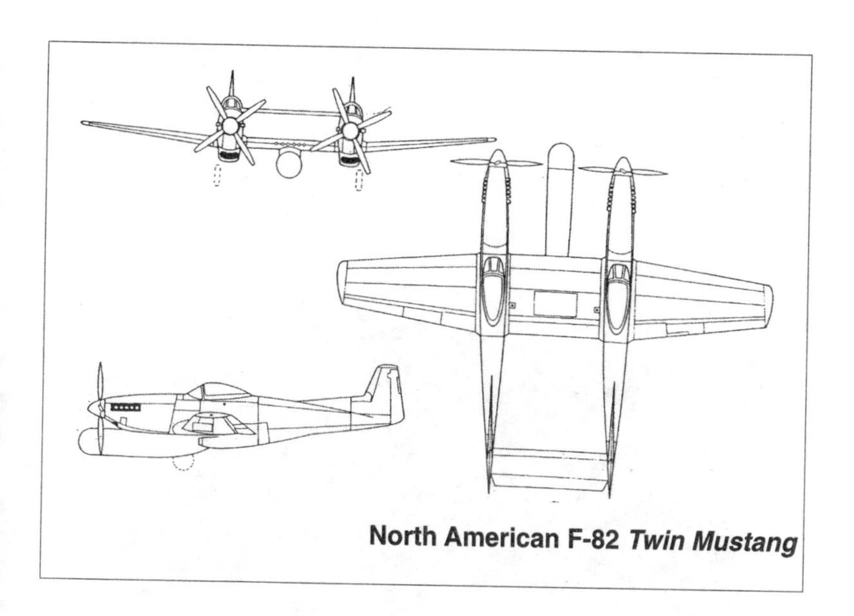

North American F-82 *Twin Mustang*

North American P-51 *Mustang*

Service: U.S. Air Force; U.S. Army Air Corps

Things to look for:

A. Air scoop located under fuselage and beneath cockpit.
B. Squared-off tail surfaces.
C. Squared-off wing tips.
D. Exhaust stacks located midway of nose, behind propeller.

Description:

The P51 was designed as a long-range, high-altitude bomber escort. Its high speed and maneuverability at high altitudes made it especially effective in high-altitude air-to-air engagements during escort missions. Further, if needed, its high speed allowed it to run away from most opponents.

The P-51 *Mustang* first flew 26 October 1940. The British Purchasing Commission (BPC) approached North American Aviation to build the Curtiss P-40 *Warhawk*. North American told the BPC that they could design and build a better fighter in the same amount of time it would take to tool up to build the P-40. The XP-51 rolled out of the North American plant doors, 127 days later. The rest is history. The P-51D introduced the bubble canopy that is the hallmark of the *Mustang* line. A total of 9603 P-51Ds were built—more than any other P-51 model. Today more than 100 P-51s survive, including the XP-51!

Specifications:

Length	32 feet, 3 inches
Wing Span	37 feet
Height	8 feet, 8 inches
Crew	1
Maximum Speed	437 mph
Service Ceiling	42,000 feet
Empty Weight	7125 pounds
Maximum Weight	9450 pounds
Engine	Rolls Royce Merlin V-1650, 12-cylinder in-line engine

Armament: Six 0.5-inch machine guns and ten 5-inch rockets, or 2000 pounds of bombs

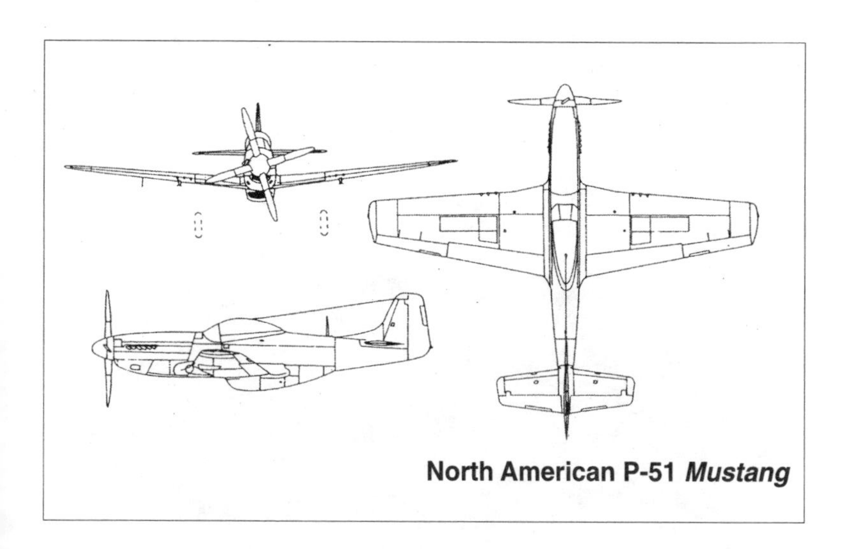

North American P-51 *Mustang*

Boeing/Stearman PT-17/N2S *Kaydet*

Service: U.S. Air Force; U.S. Navy

Things to look for:

A. Single-leg landing gear.
B. Top wing is forward of the bottom wing.
C. The back of the cockpit openings are cut straight down.
D. "N" struts between upper and lower wings.
E. Top and bottom wing have the same span; front cockpit is under top wings.

Description:

The PT-17 *Kayde*t was the most widely used primary military trainer aircraft by U.S. and Allied air forces during World War II. It was easy to fly and new pilots found it to be especially forgiving at stall speeds.

The PT-17 *Kaydet* began life as the PT-9 primary trainer of 1933. Between 1936 and 1945, more than 10,000 *Kaydets* were built and delivered to American and foreign customers. Several different designations were given the *Kaydets* depending on the type of engine used. The PT-13 had a Lycoming engine, the PT-17 had a Continental engine, and the PT-18 had a Jacobs engine. The PT-27, built for Canada, was equipped with an enclosed cockpit and different instruments. The slow, low-level flying capabilities of the *Kaydet* made it an excellent platform for crop dusting. Several *Kaydets* still survive and are seen at airshows as aerobatic aircraft. One has been modified by adding a jet engine under the fuselage!

Specifications:

Length	25 feet, 0.25 inches
Wing Span	32 feet, 2 inches
Height	9 feet, 2 inches
Crew	2
Maximum Speed	124 mph
Service Ceiling	11,200 feet
Empty Weight	19,367 pounds
Maximum Speed	2717 pounds
Engine	Continental R670 radial engine, 220 hp
Armament	None

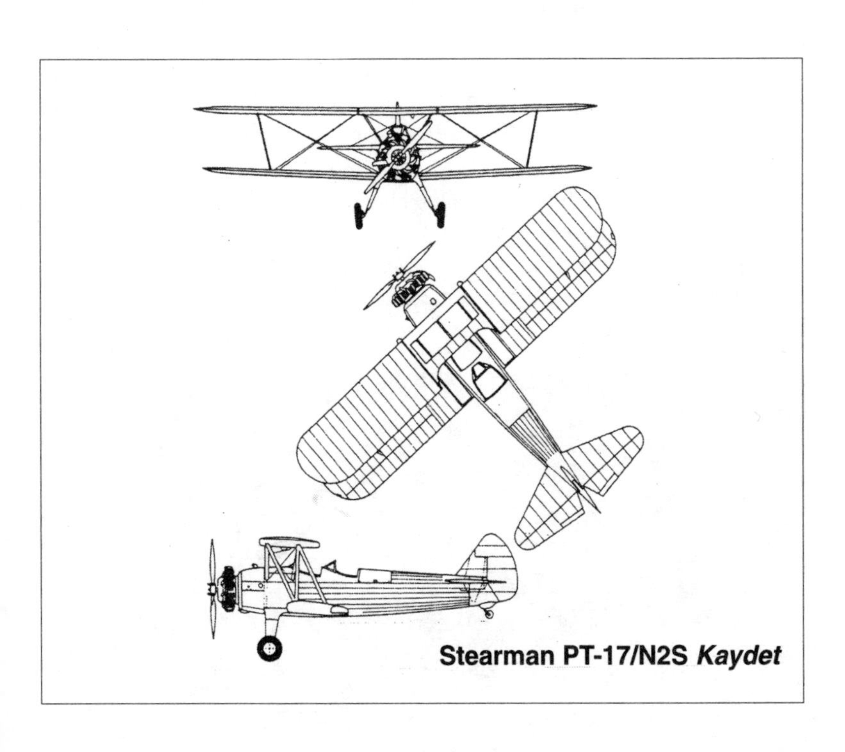

Stearman PT-17/N2S *Kaydet*

North American T-28 *Trojan*

Service: U.S. Air Force, U.S. Navy

Things to look for:

A. Wings mounted low on fuselage.
B. Tricycle landing gear.
C. Tandem cockpit.
D. Air scoop located on top of engine cowling.
E. Air scoop located on left (port) side of engine cowling.

Description:

This two-seat aircraft was the first trainer to be equipped with tricycle landing gear and a steerable nose gear. With these, good speed characteristics, and a cockpit design similar to early jets, the T-28 provided the trainee with a straightforward approach to the jet trainers he would soon be flying.

The T-28 *Trojan* first flew on 26 September 1949. Originally designed for the U.S. Navy to replace the SNJ, it did not win a production contract. The USAF was also looking for a replacement for its AT-6 *Texans*. The design was modified to meet the USAF requirements and became the T-28 *Trojan*. The *Trojan* was finally accepted by the U.S. Navy in 1954 and more than 489 units were built for the Navy between 1954 and 1955. The Navy started phasing out *Trojans* in 1984. *Trojans* were extensively used in carrier qualification from the mid-1950s to the mid-1980s. T-28s can still be seen at airports all over southern California.

Specifications:

Length	33 feet
Wing Span	40 feet, 1 inch
Height	12 feet, 8 inches
Crew	2
Maximum Speed	343 mph
Service Ceiling	37,000 feet
Empty Weight	6424 pounds
Maximum Weight	8486 pounds
Engine	Wright R-1820, 9-cylinder radial engine, 1424 hp
Armament	None

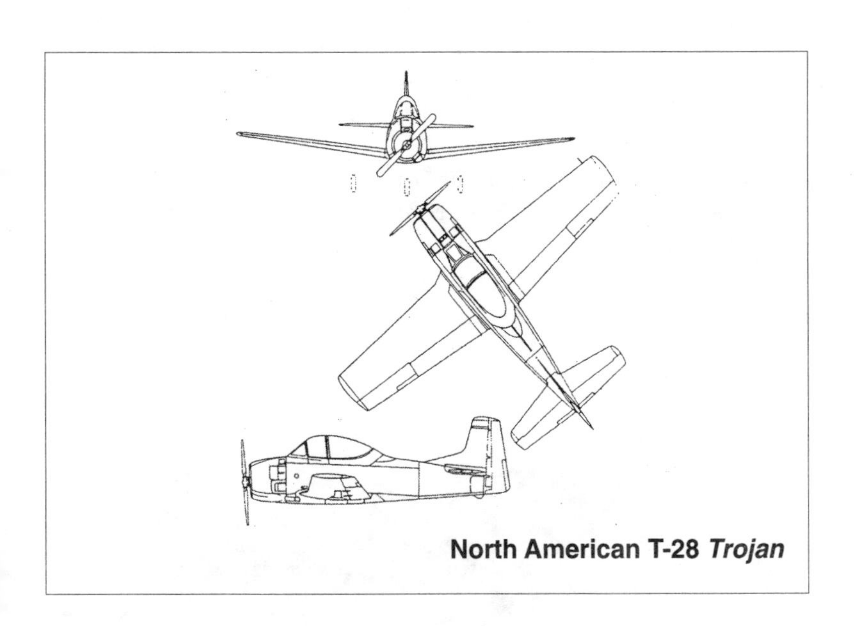

North American T-28 *Trojan*

Grumman TBF/TBM *Avenger*

Service: U.S. Navy, U.S. Marine Corps

Things to look for:

A. Wings mounted midway on fuselage.
B. Tandem cockpit.
C. Long cockpit canopy.
D. Most models have a gun turret located at the rear of the canopy.
E. A notch on the bottom of the fuselage near the tail, just in front of the tail wheel.

Description:

The TBF, a mid-wing monoplane, was designed to meet a Navy requirement for an aircraft which could carry one torpedo or three 500-pound bombs, with a three-man crew and powered gun turret. The torpedo and/or bombs were to be carried in an internal bomb bay. The aircraft's primary mission was attacking surface vessels or shore targets. At 18,000 pounds fully loaded, the TBF was the heaviest of the Navy's carrier-based aircraft. The TBF would replace the Douglas TBD *Devastator*.

The Grumman TBF *Avenger* first flew on 7 August 1941. After production began, Grumman contracted with General Motors to build the *Avenger*, to be designated as the TBM. After successful World War II usage, the *Avenger* served with the Navy until the mid-1950s. After phase-out, several TBMs became fire bombers.

Specifications:

Length	40 feet, 11.5 inches
Wing Span	54 feet, 2 inches
Height	16 feet, 5 inches
Crew	3
Maximum Speed	267 mph
Service Ceiling	23,400 feet
Empty Weight	10,843 pounds
Maximum Weight	18,250 pounds
Engine	Wright R2600 radial engine, 1900 hp

Armament: Three 0.5-inch machine guns and one 0.3-inch machine gun; one 2000-pound rated bomb load in lower fuselage weapons bay, two 250-pound bombs on underwing hardpoints, or one 22-inch Mk 13-2 torpedo in weapons bay instead of bomb(s)

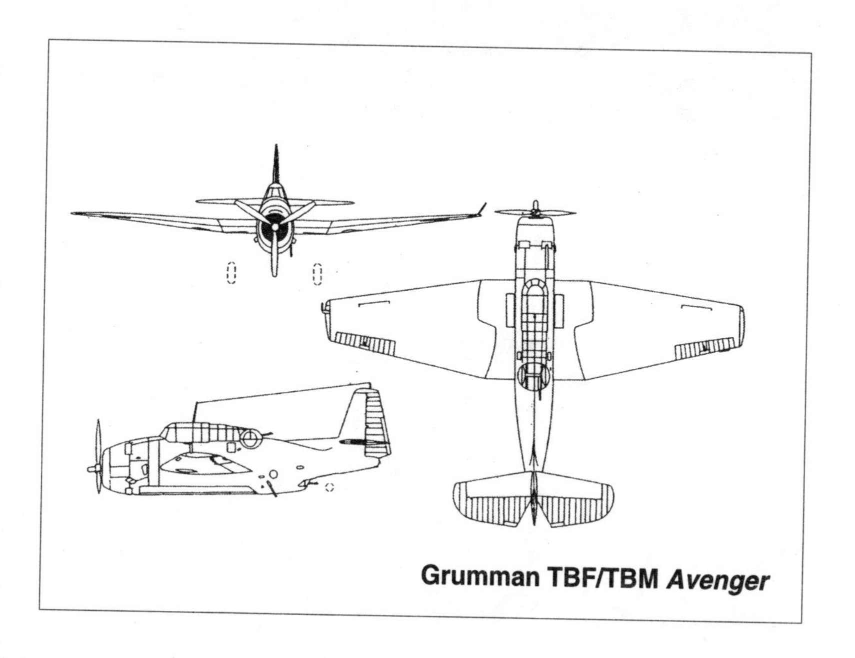

Grumman TBF/TBM *Avenger*

Section 10
New Aircraft

Lockheed/Martin F-22 *Raptor*

Service: U.S. Air Force

Things to look for:

A. Twin tails canted outboard.
B. Triangular-shaped tails.
C. Rectangular engine intakes located under cockpit.
D. Bubble canopy.

Description:

The F-22 *Raptor* is a replacement for the F-15 *Eagle* air-superiority fighter. It combines stealth design with the supersonic, highly maneuverable, dual-engine, long-range requirements of air-to-air fighter and it also will have an inherent air-to-ground capability, if needed.

The F-22 *Raptor* first flew on 7 September 1991. The USAF selected the Lockheed/Martin F-22 over the Northrop YF-23 in 1991 as a replacement for the McDonnell Douglas F-15 *Eagle*. The Air Force plans on buying 339 of the *Raptor*, with the first *Raptors* going to Tyndall AFB, Florida, during the fall of 2002. The two Pratt and Whitney F-119 low-bypass turbofans rated at 35,000 pounds of thrust will enable the *Raptor* to cruise at Mach speeds without the use of after burner. The engines also have vectoring exhaust to increase the pitch and turn rate. The F-22 *Raptor* is Lockheed's first fighter since the F-104 *Starfighter* of the 1950s.

Photo by Ray Rivard

Specifications:

Length	64 feet, 2 inches
Wing Span	43 feet
Height	17 feet, 7 inches
Crew	1
Maximum Speed	Mach 2-plus (estimate)
Service Ceiling	50,000 feet (estimate)
Empty Weight	30,000 pounds (estimate)
Maximum Take-off Weight	60,000 pounds (estimate)

Armament: The F-22 is capable of carrying existing and planned air-to-air weapons. These include a full complement of medium-range missiles such as the advanced medium-range air-to-air missile (AMRAAM), and short-range missiles such as the *Sidewinder*. The F-22 also will have a modernized version of the proven M61 internal gun and growth provisions for other weapons. The aircraft is also capable of carrying Joint Direct Attack Munitions (JDAMs) and other ground-attack weapons.

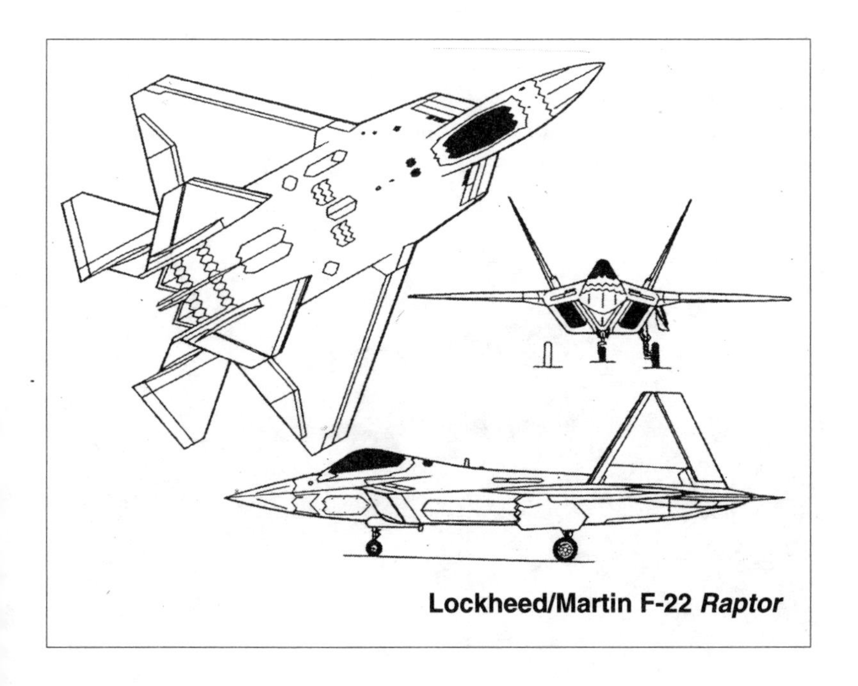

Lockheed/Martin F-22 *Raptor*

Bell-Textron MV-22 *Osprey*

Service: All U.S. military branches

Things to look for:

A. Huge propellers located at the ends of the wing.
B. Twin tail.
C. Main landing gear located in pods located on side of fuselage.
D. Wing mounted high on fuselage.
E. Lower rear surface of fuselage slants upward.

Description:

The primary function of the Bell-Textron MV-22 *Osprey* is to act as an amphibious assault transport of troops, equipment, and supplies from assault ships and land bases. As such the MV-22 is a multi-engine, dual-piloted, self-deployable, medium-lift, Vertical Take Off and Landing (VTOL) tiltrotor aircraft designed for combat, combat support, combat service support, and Special Operations missions worldwide.

The MV-22 *Osprey* first flew in March 1989. Bell Helicopter and Textron-Vertol teamed during the 1980s to develop the MV-22 as a replacement for the aged fleet of CH-46 *Seaknight* and CH-53D *Sea Stallion* medium-lift helicopters. The MV-22 entered the U.S. Marine Corps inventory at a Low Rate Initial Program (LRIP) during 1997 and completed more than 300 carrier landings.

Specifications:

Length	57 feet, 4 inches; 62 feet, 7 inches (rotor folded)
Rotor Diameter	38 feet
Height	21 feet, 9 inches
Crew	3
Maximum Speed	345 mph
Service Ceiling	26,000 feet
Empty Weight	33,140 pounds
Maximum Weight, Vertical Take-off	52,870 pounds
Maximum Weight, Short Take-off	57,000 pounds

Armament: The aircraft will be equipped with a 12.7 mm (0.5-inch) tur-reted gun system, which will be supplied by General Dynamics. Additional weaponry will evolve as the program progresses.

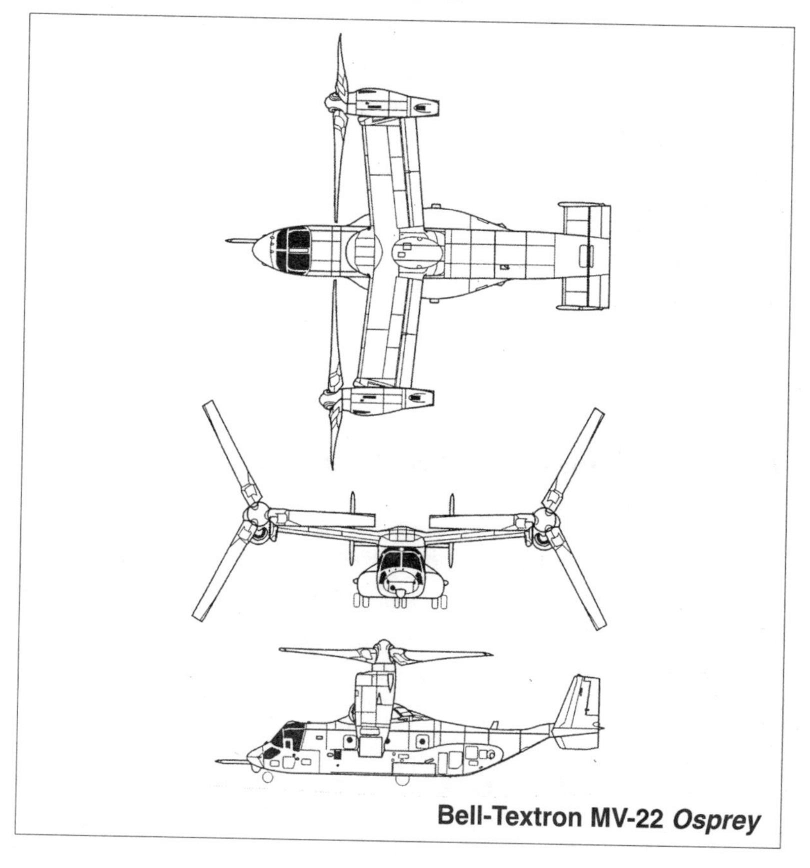

Bell-Textron MV-22 *Osprey*

Appendix
Aviation Museums in Southern California

(All information is subject to change without notice. Please contact before visiting. Please advise publisher of changes.)

Air Force Flight Test Center Air Museum

405 South Rosamond Boulevard
Edwards AFB, CA 93524-8215
Hours: 9 a.m. to 5 p.m., Tuesday through Saturday
Admission: Free
Phone: (310) 392-8822
Website: http://afftc.edwards.af.mil/museum/index.html

Classic Rotors

Ramona Airport, Ramona, CA
c/o Mark DiCiero
6161 El Cajon Boulevard
San Diego, CA
Hours: Phone for information
Admission: Phone for information
Phone: (619) 427-1330 (Voice Mail)
Website: www.rotors.org

Flying Leatherneck Aviation Museum

Marine Corps Air Station Miramar
P.O. Box 45316
San Diego, CA 92145-0316
Hours: Monday through Saturday, 9 a.m. to 3 p.m.
Admission: Free
Phone: (858) 693-1723
Website: www.flyingleathernecks.org

March Field Air Museum

P.O. Box 6463
March AFB
Riverside, CA 92518
Hours: 9 a.m. to 4 p.m., Daily
Admission—suggested donation:
Adults $5, Kids $2, Family (two adults and minor kids) $10
Phone: (909) 697-6602
Website: www.marchfield.org

Museum Of Flying

2772 Donald Douglas Loop North
Santa Monica, CA 90405
Hours: 10 a.m. to 5 p.m., Saturday and Sunday only
Admission: Seniors $6, Adults $8, Ages 3 to 17 $4
Phone: (310) 392-8822
Website: www.museumofflying.com

Palm Springs Air Museum

745 North Gene Autry Trail
Palm Springs, CA 92262
Hours: 10 a.m. to 5 p.m., Daily (closed Thanksgiving and Christmas)
Phone: (760) 778-6262 ext. 222
Website: www.air-museum.org

Planes of Fame Air Museum

Chino Airport
7000 Merrill Avenue
Chino, CA 91710
Hours: 9 a.m. to 5 p.m., Daily (closed Thanksgiving and Christmas)
Admission: Adults $8.95, Ages 5 to 12 $1.95
Phone: (909) 597-3727
Website: www.planesoffame.org

San Diego Aerospace Museum

2001 Pan American Plaza
Balboa Park
San Diego, CA 92101
Hours: 10 a.m. to 4:30 p.m., Daily (doors close at 4 p.m.)
Summer Hours (Memorial Day through Labor Day):
10 a.m. to 5:30 p.m. (doors close at 5 p.m.)
Admission: Adults $8, Seniors (Ages 65+) $6,
Ages 6–17 $3, Groups of 10 or more $6,
Active Duty Military and Ages under 6 free
Phone: (619) 234-8291
Website: www.aerospacemuseum.org

San Diego Aircraft Carrier Museum (USS *Midway*)

(Midway Museum Office)
1355 North Harbor Drive
San Diego, CA 92101
Hours/Admission: The USS *Midway* is not open at the time of this writing. Please contact the website for updates.
Phone: (619) 702-7700
Website: www.midway.org

Yanks Air Museum

Chino Airport
7000 Merrill Avenue
Chino, CA 91710
Hours: Phone first
Admission: Phone first
Phone: (909) 597-1734

Index

Sunbelt Publications

"Adventures in the Natural and Cultural History of the Californias"
General Editor—Lowell Lindsay

Southern California Series:

Title	Author
Geology Terms in English and Spanish	Aurand
Portrait of Paloma: A Novel	Crosby
Orange County: A Photographic Collection	Hemphill
California's El Camino Real and its Historic Bells	Kurillo
Mission Memoirs: Reflections on California's Past	Ruscin
Warbird Watcher's Guide to the Southern California Skies	Smith
Campgrounds of Santa Barbara and Ventura Counties	Tyler
Campgrounds of Los Angeles and Orange Counties	Tyler

California Desert Series:

Title	Author
Anza-Borrego A to Z: People, Places, and Things	D. Lindsay
The Anza-Borrego Desert Region (Wilderness Press)	L. and D. Lindsay
Geology of the Imperial/Mexicali Valleys (SDAG 1998)	L. Lindsay, ed.
Palm Springs Oasis: A Photographic Essay	Lawson
Desert Lore of Southern California, 2nd Ed.	Pepper
Peaks, Palms, and Picnics: Journeys in Coachella Valley	Pyle
Geology of Anza-Borrego: Edge of Creation	Remeika, Lindsay
California Desert Miracle: Parks and Wilderness	Wheat

Baja California Series:

Title	Author
The Other Side: Journeys in Baja California	Botello
Cave Paintings of Baja California, Rev. Ed.	Crosby
Backroad Baja: The Central Region	Higginbotham
Lost Cabos: The Way it Was (Lost Cabos Press)	Jackson
Journey with a Baja Burro	Mackintosh
Houses of Los Cabos (Amaroma)	Martinez, ed.
Baja Legends: Historic Characters, Events, Locations	Niemann
Loreto, Baja California: First Capital (Tio Press)	O'Neil
Sea of Cortez Review	Redmond

San Diego Series:

Title	Author
Rise and Fall of San Diego: 150 Million Years	Abbott
Only in America	Alessio
More Adventures with Kids in San Diego	Botello, Paxton
Geology of San Diego: Journeys Through Time	Clifford, Bergen, Spear
Cycling San Diego, 3rd Edition	Copp, Schad
A Good Camp: Gold Mines of Julian and the Cuyamacas	Fetzer
San Diego Mountain Bike Guide	Greenstadt
San Diego Specters: Ghosts, Poltergeists, Tales	Lamb
San Diego Padres, 1969-2002: A Complete History (Big League Press)	Papucci
San Diego: An Introduction to the Region (3rd Ed.)	Pryde
Campgrounds of San Diego County	Tyler